CLIMATE:
THE GREAT DELUSION

CLIMATE:
THE GREAT DELUSION

A Study of the Climatic, Economic and Political Unrealities

Christian Gerondeau

STACEY
INTERNATIONAL

Climate: The Great Delusion

STACEY INTERNATIONAL

128 Kensington Church Street

London W8 4BH

Tel: +44 (0)20 7221 7166; Fax: +44 (0)20 7792 9288

Email: info@stacey-international.co.uk

www.stacey-international.co.uk

ISBN: 978 1 906768 41 6

CIP Data: A catalogue record for this book is available from the British Library

© Christian Gerondeau 2010

1 3 5 7 9 0 8 6 4 2

Printed in Turkey by Mega Basim

Contents

List of Figures

Foreword

This is a most timely book. In the wake of the fiasco of the December 2009 United Nations climate change conference, intended to secure a global agreement on worldwide decarbonisation, which was followed by evidence of scandalous shortcomings in the IPCC process on which national governments were expected to base their climate change policies, there is now a greater readiness than before to question the conventional wisdom of what has become almost a new religion.

Christian Gerondeau, a distinguished French engineer, explains clearly why the conventional approach is so deeply flawed, and what needs to be put in its place. As an engineer (and he points out that one reason why China rejected the decarbonisation agenda at Copenhagen may be that its leaders are nearly all engineers by training) he is neither overcome by the cultural pessimism of the Western political classes nor seduced by their grandiose oratory. He keeps his feet on the ground and is concerned with what works.

Thus we should focus less on the uncertain science of global warming and on highly speculative horror stories about its possible impact, and more on the practical way forward, where the key is technological development, and the enhanced opportunities which that provides to adapt to any climate change that may occur. Moreover, while the technologically inept and economically ruinous decarbonisation route, even if it could be globally agreed (which it cannot), would intensify the problem of world poverty, the way ahead which Christian Gerondeau charts would alleviate it.

As Gerondeau observes, future generations will find it hard to believe the story he accurately recounts in this book. It is a book that deserves to be widely read.

Nigel Lawson, April 2010

A sceptical note

The great advantage of a scientific mind is that it reasons in an objective and inventive manner. This is true of the renowned engineer Christian Gerondeau, who has given much thought to one of the dominant concerns of our times, climate change.

Mr Gerondeau begins with an original line of thought, leading to a series of fundamental questions. He believes that we are on the wrong track since it is not realistic to try to reduce carbon dioxide emissions. In his view, the only way of doing so would be to stop extracting fossil energy resources or at least to set a ceiling on them. However, we are doing exactly the opposite by encouraging oil exploration, developing natural gas transport networks and by opening new coalmines. Once these fossil energies have been extracted, they will naturally be consumed and will produce carbon dioxide emissions. If we do not extract them, other countries will, thus producing the same volume of carbon dioxide emissions. The only solution would be to limit the extraction of oil, natural gas and coal, but there is no global consensus to propose this.

Christian Gerondeau puts forward another original idea, indicating that there will probably be little harm done to the planet since he shows, as have other scientists, that no definite relationship has been established between the atmospheric concentration of carbon dioxide and possible changes in the climate. The latter are due to far more complex causes which probably trace their origins well beyond the atmosphere.

Mr Gerondeau also deduces that the formidable sums of money that various countries throughout the world are committed to devoting to 'saving the planet' will be completely wasted, such as those serving to cover France with wind turbines, which disfigure our landscapes and for which we have no need in terms of energy supply. The result is poor use of public funds for an ineffective production of electricity.

This book has the virtue of helping us question generally accepted ideas and perhaps discover more realistic solutions for the future.

Valéry Giscard d'Estaing

Some key questions

Question 1

Is it possible to imagine that mankind would leave any oil, natural gas or coal that could be economically exploited in the planet's underground?

Answer

The answer is no. The available reserves are not infinite and humanity will look to exploit the world's oil deposits, gas fields and coal mines to the very last drop, cubic metre and tonne economically reachable.

Question 2

Is it possible to stop mankind's use of oil, natural gas and coal resulting in the release of CO_2 when burned?

Answer

The answer is also no for both physical and economic reasons. In most cases there are no technical means of preventing the CO_2 produced being released into the atmosphere. And if a technical solution for sequestrating the CO_2 produced in electric power plants or factories was one day perfected, the very high cost would prevent it playing a significant role at a planetary level.

Question 3

Is it realistic to try to prevent the continued increase in the concentration of CO_2 in the earth's atmosphere?

Answer

Once again the answer is no. In the main, CO_2 emissions are the consequence of the use of fossil hydrocarbons, and mankind's needs are such that it will use all the resources economically available, as they are vital to its development and to bring humanity out of poverty. What developed countries do not use will be used by China, India and other nations. The official wish to halve emissions by 2050 in relation to the current level, i.e. to divide them by four in relation to the present trend, is just wishful thinking. The truth is we can do nothing about it.

The conclusion is clear: the quantities of hydrocarbons still present underground, and nothing else, will determine the amount of CO_2 to be released in the atmosphere in the coming decades. Simple reckonings then show we have to expect the amount of CO_2 lying over our heads to about double before the end of this century.

Question 4

Will CO_2 emissions continue to grow indefinitely?

Answer

As most of the CO_2 emissions come from the combustion of fossil hydrocarbons and as, starting with oil, these will gradually be exhausted, the answer is once again no. There will come a time when energy-related emissions will hit a ceiling and start to decrease before almost disappearing at the end of the twenty-first century when the reserves are exhausted.

Question 5

Are we heading for disaster?

Answer

Fortunately there is every chance this is not the case. In the distant past, the amount of CO_2 in the atmosphere has been sometimes five to twenty times what it is today, and life on earth was not harmed. No serious examination backs up the apocalyptic predictions we are fed daily about the consequences of the inevitable increase in the CO_2 concentration in our atmosphere. We are told so many obvious falsehoods regarding, for instance, the Himalayan glaciers or the effect of rising sea levels, which have no basis in scientific truth that it is hard to trust those who express them.

I   We can do nothing about it

The climate is changing. Glaciers are melting. Polar bears are dying out. We must take urgent action to avoid disaster on a world scale. We must all join forces and make a contribution. It is high time to change our behaviour and save the planet. Are we going to sacrifice our children's futures because of our selfishness? Our responsibility is all the greater because the solution lies in our own hands. The utmost priority is to reduce CO_2 emissions.

After being repeated a thousand times over, messages such as these have long become universally accepted in developed countries, particularly in Europe.

The advocates of these ideas were hailed for their wisdom by our contemporaries and by world leaders, to the point that they have inspired the world's most important bodies. The fight against climate change now appears at the top of the agenda of every meeting of leaders, in the European Union and on the world level.

In this context, Al Gore, former vice president of the United States and author of the film *An Inconvenient Truth*, and the United Nations' Intergovernmental Panel on Climate Change (IPCC) were jointly awarded the Nobel Peace Prize in Oslo on 10 December 2007.

At the Hokkaido Summit held on 8 July 2008 in Japan, the leaders of the G8 countries (Canada, France, Germany, Italy, Japan, Russia, United Kingdom and United States) adopted the target of reducing global carbon dioxide (CO_2) emissions by at least 50 per cent by 2050. The following day, the largest developing countries, which then became known as the G5

(Brazil, China, India, Mexico and South Africa), expressed for the first time a 'shared vision' for concerted action to 'reduce emissions in the long term'.

The cause seemed to be settled.

However, the truth of the matter is that the cause was not settled at all! We were the victims of a global myth. This has already happened in the past: our forefathers were convinced that the earth was the centre of the universe and were prepared to send anyone who believed otherwise to the stake.

The subject has changed, but the passions that surround it have not. Rarely has a debate preoccupied opinion and the media as much as the one that is raging today over climate change and humankind's possible responsibility in this regard.

The climatologists themselves are divided into two opposing factions. Most appear to side with the 'official' camp, which supports a point of view that can be summarized in four points:

- The temperature of the earth is rising and will continue to do so to a dangerous degree;
- Human activities, especially emissions of carbon dioxide (CO_2), are the major cause of this phenomenon;
- We therefore need to take action to control and reduce these emissions;
- The means to achieve this exist and must be urgently implemented to 'save the planet'.

But other climatologists, who are increasingly making their voices heard, are challenging the very foundations of this reasoning. According to them, climate variations have always existed and nothing proves that human activities, notably CO_2 emissions, have a significant influence on such changes and on the average temperature of the earth.

When one is not a climatologist, of whom there are precious few in the world, is it possible to form an opinion? Faced with the difficult nature of the subject and the extent of the controversy, there is a strong temptation to answer this question in the negative. The present book will endeavour to refute this and prove that a rational approach, that is to say starting with the facts, allows us to reach a number of conclusions that are difficult to dispute and to look at this complex issue in a new light.

Instead of jumping into a debate about climate change and the possible responsibility of human activities that divides climatologists among themselves, the reasoning proposed turns the problem around, starting with a simple question: 'What is our room for manoeuvre with regard to CO_2 emissions and is there anything we can do about them?'

Of course, our instinct is to respond positively when we think about what we can – or could – do to limit our energy consumption and our emissions, whether as an individual or at the level of our country. It is true that there is sometimes considerable margin for action. We assume that, if everyone behaves as best we could and if we combine our efforts, the problem can be solved.

However, this is not the right approach as it is misleading. There is another way of looking at things, which is to study how emissions from each of the main sources of greenhouse gases resulting from human activities are likely to evolve in the future. If we exclude deforestation which, although it is an important phenomenon, is by nature temporary, only four factors are responsible for three-quarters of total emissions. In terms of CO_2, which alone represents the great majority of greenhouse gases of human origin, it mainly emanates from the combustion of three products: oil, natural gas and coal. As for methane, which represents a large part of the remaining emissions, it mainly emanates from livestock farming, especially of ruminants.

If we wish to fight man-made greenhouse gas emissions, we must therefore reduce the discharges from oil, natural gas, coal and cattle. And to find out whether this objective is achievable, we must start by asking two questions: 'Is it realistic to imagine that humanity will leave any of its economically reachable reserves of oil, natural gas and coal unexploited?' and 'Is it conceivable to prevent people emerging from poverty in developing countries from adopting our Western-style diet?'

In fact, the questions almost provide the answer and help us understand why greenhouse gas emissions have never increased as quickly as at present, although there have never been so many voices speaking out against them. From 2000 to 2009, emissions have increased by an average of more than 3 per cent per year compared with 0.9 per cent per year between 1990 and 2000. Developed countries have more or less controlled their emissions, but those in the rest of the world – i.e. close to six billion people instead of about one billion in developed countries – have increased a great deal faster than anticipated.

There is no need to look any further for the source of the confusion than the 2008 G8 meeting in Japan. Admittedly, the G5 countries did agree at this time with the rich nations' position that climate change could be one of the major challenges of our time, but they would not go any further, and refused to back the target of reducing global emissions by at least half by 2050 and to commit themselves to working towards a 'carbon free' world as they were asked to. The five emerging nations refused to go beyond a declaration of principle, as long as developed countries continued to be incoherent by not committing themselves, first, to setting a target to reduce their own emissions by at least 80 per cent by 2050, as they currently pollute far more than the rest of the world in relation to their populations and, second, to reducing their emissions by at least 25 per cent by 2020, compared with 1990 levels, proof of their determination to take action. But members of G8 refused to make any such commitments.

Speaking with a single voice, the five southern countries added that, if any, the current problem stems, above all, from the developed countries' past consumption, and that it was therefore up to the latter to pay the price, and not up to them as they have a vital need for fossil fuels (oil, natural gas and coal) to ensure their development.

As an answer, the rich countries indicated that they were indeed willing to make the bulk of the efforts required to halve global emissions by 2050. Nonetheless, as global emissions are now split 50/50 between developed and emerging countries, the latter would also have to significantly restrict their emissions to reach such a global goal, although nobody in their right mind can seriously contemplate this. Given that developing countries' populations will continue to grow, that the greater part of humanity is currently poverty-stricken if not destitute and that more than one and a half billion people quite simply do not have access to electricity, you have to be either completely thoughtless or heartless to ask emerging countries to further reduce their emissions, i.e. their energy consumption, with all the dramatic consequences that this would entail for their inhabitants. Obviously, exactly the opposite will happen. Excluding China, developing countries emit 1.4 tonnes of CO_2 per inhabitant per year, to be compared with 15 tonnes on average for developed countries.

How would it be possible for them to reduce their emissions any further? And why should they be condemned to poverty forever?

These discussions left a deep sense of unease and the matter was postponed until the end of 2009, when a world climate conference was to be held in Copenhagen to decide on a new greenhouse gas emission programme to follow on from the Kyoto Protocol, which was to end in 2012. This was to be the moment of truth, as it would become clear that the targets set by the world's leading nations were impossible to achieve, first because they presuppose unrealistic efforts on the part of the developed world and second because they require the developing countries' agreement, which they were not prepared to give at any price.

So it was not difficult to forecast that Copenhagen was doomed to be a dramatic failure, as I did in the French version of this book published at the beginning of 2009. There will never be any follow-up to Kyoto.

A few basic facts are required to help understand this point. First, our planet is fortunate to have an atmosphere made up of various gases, especially water vapour, that warm it by preventing some of the solar radiation it receives from the sun going back into space. These are the famous 'greenhouse gases', without which the average temperature on the surface of the globe would be about 30 degrees celsius lower, leaving us shivering with cold, assuming we were there at all. Among them is carbon dioxide, or CO_2, whose mass in the atmosphere two centuries ago was approximately *2,000 billion tonnes*, a quantity that had probably hardly varied for thousands of years.

Things changed with the industrial era when men began large-scale exploitation of coal, followed by oil and natural gas, which had accumulated hundreds of millions of years ago in the earth's depths. CO_2 emissions from the use of these three products have now reached *30 billion tonnes* per year and continue to increase every year. The origins of these emissions are well known. 13 billion tonnes are due to coal, 11 billion to oil and the remaining 6 billion to natural gas.[1] As CO_2 has a lifespan that is usually said to exceed one hundred years in the atmosphere, it is hardly surprising that the stock surrounding the globe has considerably increased since the beginning of the industrial era. According to the reference data coming from the Mauna Loa observatory in Hawaii, it currently stands at 2,800 billion tonnes, i.e. 800 more than two centuries ago, a 40 per cent increase.

It is not only CO_2 resulting from the combustion of coal, oil and gas that must be taken into account. Some is also coming from the deforestation taking place in various parts of the world, as well as from various agricultural activities. Finally, although CO_2 is by far the most significant greenhouse gas apart from water vapour, it is not the only one. The picture is therefore not complete if we fail to take into account the other gases that contribute to the phenomenon, notably methane. The annual contribution to the greenhouse effect of sources other than oil, natural gas and coal is presently estimated at a total of approximately 20 billion tonnes of

CO_2. Consequently, total emissions of human origin into the atmosphere currently amount to *around 50 billion 'equivalent tonnes'* of CO_2 every year.

To see how the situation is likely to evolve in the future, we shall now examine the chief actors involved.

Oil

The events at the G8 Summit in Hokkaido on 7 July 2008 were a perfect illustration of the contradictions that can sometimes be found even in the most well informed minds, or those that claim to be so. In the morning, the participants anxiously pondered on how to obtain more oil and begged the oil-producing countries to increase their production in order to bring down prices. In the afternoon, in another meeting, they discussed imposing taxes to discourage the use of fossil fuels, apparently without being aware of what the international press described as a 'glorious incoherence' (*International Herald Tribune*, 10 July 2008). On one hand, they were clamouring for more oil in order to reduce prices and on the other, they were insisting that it should not be used and should be taxed to increase its price for users in order to reduce emissions! What do we really want, one thing or the other?

Actually, the answer is quite clear. Oil plays such an important role in the world economy today that it is impossible to do without it, although of course we all know it will run out one day. With the exception of the relatively small amount required for the chemicals industry – particularly for making plastics – oil is above all used today to ensure the physical exchanges inherent to all economic and human activities.

All air and marine transport and more than nine-tenths of land transport would be impossible without liquid fuel, which today means oil products. This is quite obvious for aeroplanes and ships, but if we think about it carefully, it is just as true for the great majority of people's day-to-day trips and for land-based freight transport, given that modes of transport are not communicating vessels. How can we supply our shops, supermarkets, most of our workshops and even our factories other than by lorry or van? Who can seriously imagine that it can be done by train or barge? In our cities, how can we travel conveniently from one suburb to another except in an individual vehicle, given the low population densities generally found outside the town centres that only account for a minority of the population and the jobs?

As proof of the fact that there are actually very few possibilities for substituting one means of transport for another, it is sufficient to note that the very sharp rise in fuel prices at petrol stations up to the middle of 2008 hardly changed people's behaviour at all. In the United States, where prices had risen in three years from 1.3 to more than 4 dollars per gallon (3.8 litres), i.e. a 200 per cent increase, traffic volume only fell by 3 per cent! It is difficult to imagine lower short-term elasticity.

In Europe, where the relative increase was more moderate due to the heavy taxes, the same phenomenon occurred. Contrary to the impression given by the media of cases of people giving up cars for public transport or bicycles, nothing really changed and the volume of traffic was not reduced significantly. For the 90 per cent of Europeans who do not live or work in the centre of large towns, cars are indispensable for getting to work, doing the shopping, visiting friends and family, practicing a sport and going away for the weekend or on holiday. If necessary, other items of expenditure are cut to compensate for increases in fuel costs, but people go on using their cars because they are an intrinsic part of their everyday lifestyle, if not of life itself, as a result of changes in the geography of housing and other land uses.

Doubtless because they usually live in town centres, people who advocate a 'change in behaviour' and are convinced that this is possible, actually know very little about how the very great majority of the inhabitants live in all developed countries. By telling them repeatedly that they should change their lifestyles and use their cars less – which is simply impossible – all they manage to do is to make them feel guilty.

The Chinese and the Indians have no such scruples. For instance, the recent 'Chinese miracle' was based on two pillars: first, the construction, at the frenetic rate of at least one per week, of coal-fired electric power plants with the capacity of a nuclear unit and second, the creation *ex nihilo* of a motorway network that will soon surpass that of the United States. Whereas 20 years ago, China *did not have a single kilometre* of motorway, it now has 50,000 and is building 5,000 more per year as can be seen from the country's road maps, which are never up to date. Beijing is in the process of finishing its seventh ring motorway whereas Paris has still not completed its second, 40 years after the works began. As for India, it is beginning to follow China's example, with a time-lag of one or two decades. It is therefore pointless to try to convince the Chinese, the Indians and others to give up lorries and cars, as they are clearly counting on them for the greater part of their internal transportation needs, and so for their prosperity.

As for economic and human exchanges between countries and continents, what would become of them without aeroplanes and ships?

These considerations inevitably lead to a simple fact: the world's oil resources accumulated during previous eras will be used to the last drop to enable the exchanges that are vital to the life of the planet. Can anyone seriously doubt that? The frenzy of exploratory drilling following the surge in the price per barrel in 2007, proves that all economically reachable conventional reserves will be exploited one day, even if they are in very deep sea or under the poles in terrains that until recently were considered inaccessible. Similarly, existing oilfields will be exploited still further, given that on average two-thirds of their reserves are left underground today and that new extraction techniques are available which may increase this proportion.

In addition, there are other, non-conventional, oil resources. Considerable reserves exist in the form of oil shale and oil sands, notably in Canada and Venezuela; they can be profitably exploited when oil prices reach about 80 dollars per barrel. It is therefore no surprise that the major oil companies have rushed into these 'new oils' that can be transformed into fuel relatively easily, although admittedly after operations which emit large quantities of CO_2.

The drop in oil prices in the second half of 2008 has brought many projects to a standstill, but they will be resumed as soon as the market turns.

The obvious conclusion is that oil products are so useful for transporting people and goods that all the available resources will be used. And this will inevitably produce greenhouse gas emissions: there are no suitable techniques to avoid this happening, since the fuel is by nature used when the vehicles are moving, meaning that the emissions can only go into the atmosphere. Almost all oil will be used for transport and the carbon it contains will be transformed into CO_2 and could, to a greater or lesser extent, contribute to the greenhouse effect.

This obviously does not mean that efforts should not be made to economise 'black gold', quite the contrary. There are many reasons why it is essential to do everything possible to rein in our consumption of oil products, as long as this can be done at reasonable cost. The less we consume, the lower our energy bills will be, although this objective does not justify limitless expenditure to save oil. There will also be less tension on the world market and the prices will tend to increase less, although the law of supply and demand works far from perfectly when it comes to oil. The less we consume, the easier it will also be to prepare for the post-oil era, as we shall see in a later chapter. Finally, the less we consume, the more can be left for developing countries that have a vital need for oil to help fight poverty and improve the fate of their people.

However, all this will have no effect on global oil consumption or related greenhouse gas emissions. The resources which are not used by developed economies will be used by the Chinese, the Indians and other countries instead. Whatever changes may take place in the economic situation, it is supply which limits emissions and not demand.

It is therefore relatively simple to evaluate the volume of emissions that will be produced in the decades to come as a result of using up the world's remaining economically recoverable oil resources.

Reference publications (*BP Statistical Review*) estimate that today's proven oil reserves represent 40 years of production at the present rate, which slightly exceeds four billion tonnes per year. If yet undiscovered reserves are added, together with non-conventional oil shale and oil sands not yet taken into account, it follows that the total economically reachable oil reserves may represent the equivalent of close to one hundred years of production at the present rate.

Given that the annual combustion of the approximately four billion tonnes of oil produces eleven billion tonnes of CO_2, it is easy to work out that using all the available oil results in discharging in the order of 1,000 billion tonnes of CO_2 into the atmosphere.

In addition, most of these emissions will probably take place before 2050. Certain experts, proponents of the 'peak oil' theory, believe that oil production has already reached a peak and will now fall off gradually. Others predict that the production peak will not be reached for two or three decades, some forecasting even later than that. However, there is unanimous agreement that oil is not eternal and that the greater part of remaining economically recoverable resources will be consumed in the coming few decades. In truth, in terms of CO_2 emissions, the uncertainty about the exact rate is of no great consequence, as it is usually agreed CO_2 has a lifespan in the atmosphere of about one hundred years. Whether it is emitted sooner or a little later makes no significant difference to the stock of CO_2 floating above our heads. *The main thing is to understand that all the economically reachable oil accumulated over the past hundreds of millions of years will be exhausted in the coming few decades and that the inevitable result will be the emission of approximately one thousand billion tonnes of CO_2 into the Earth's atmosphere.*

People today have a curious attitude to this issue. They are absolutely certain of two things: first, that oil reserves will soon run out and second, that we must reduce emissions resulting from using oil. They fail to see that these two statements are contradictory. If the reserves run out, this means that none of the planet's reachable oil resources have been left unexploited, whereas the only way of reducing emissions would be precisely that: to leave them unexploited.

Is the situation any different for natural gas and coal?

Natural gas

Natural gas is almost as precious as oil, even if it does not have the same uses. Roughly half is used for heating buildings and for cooking purposes and another large percentage is used to make electricity.

It is still used relatively little for transport, which currently only accounts for five per cent of total use. Nonetheless, there are already road vehicles that run on natural gas. For example, some of the buses in large cities such as Paris run on natural gas without anyone noticing, and have the reputation of being environmentally friendly as they are considered to generate less pollution. Some cars are now also available but it has to be said that, in both cases, natural gas vehicles are not competitive under the present conditions. They require expensive equipment to compress the gas and their range is relatively limited. Also, whatever guarantees may be given by the manufacturers, the thought of travelling with a tank full of compressed gas may somewhat dampen people's enthusiasm. It is also possible, albeit more costly, to transform natural gas as a substitute to oil, by a process known as GTL (gas to liquid). When there is no more oil, natural gas can take over.

Today, with the exception of a small percentage devoted to non-energy uses (i.e. chemicals), almost all the natural gas used every year is currently burnt, as in the case of oil, and this combustion also gives rise to CO_2 emissions which may contribute to the greenhouse effect. This being said, natural gas emits less from this point of view than coal. For the same amount of heat produced, it discharges less than half the amount of CO_2 than coal, which is why producing electricity in gas-fired power plants is a relative improvement on coal-fired plants from this point of view.

Be that as it may, every year natural gas is currently responsible for the emission of about 6 *billion tonnes* of CO_2 into the atmosphere, which represents 20 per cent of the thirty billion tonnes resulting from the combustion of the three types of hydrocarbons, i.e. coal, oil and natural gas.

It is possible that natural gas will play a greater role in transport in the future as the oil reserves run out and there is no way of preventing the emission of CO_2 into the atmosphere for this type of use (combustion). The same can be said of the domestic use of natural gas.

Electricity production is the main use of natural gas for which it may one day be technically possible to prevent the release of CO_2, produced by its combustion into the atmosphere. But this possibility implies that techniques for capturing and storing the power plants' emissions are developed and put into practice, and that the considerable additional costs can be met. Known under the initials CCS (Carbon Capture and Storage), these techniques are still at the prototype stage and mainly being considered for coal-fired plants as we shall see later in the chapter.

In fact, the conclusion is the same for natural gas as for oil: it is such a precious product that it is impossible to imagine that humanity will not exploit all the world's resources. The total reserves will be used up, as and when new fields are discovered and new means of transport are developed to transport the gas to where it is to be used. Furthermore, practically all its uses will release CO_2 into the atmosphere.

According to current estimates, proven reserves of conventional natural gas represent 60 years at the present rate of consumption. However, as with oil, it is accepted that the proven reserves do not represent all the resources inherited from past geological eras, and it would be reasonable to estimate the real conventional reserves at the equivalent of at least one hundred years at the present rate of consumption, therefore producing *in the region of 600 billion tonnes* of CO_2. If the CO_2 capture and sequestration techniques are perfected and implemented one day, they will barely reduce this figure as they could only be applied to a minority of emissions and then only in several decades time. At best, the emissions will be reduced by a few tens of billions of tonnes.

What is more, similarly to oil, most of this consumption will probably take place during a short period, exclusively in the present century. The rate of natural gas use is increasing rapidly every year, as world consumption rose from 2,245 billion cubic metres in 1997 to 2,921 in 2007 – an increase of 30 per cent in just ten years.[2]

Coal

Although it already ranks first in terms of annual atmospheric emissions, coal will increasingly be the major source of emissions. Coal consumption is developing more rapidly than any other and it has the largest reserves of all the fossil hydrocarbons.

In 1973, the world production of coal was three billion tonnes whereas it now amounts to six. The acceleration has been especially sharp since 2000 when it was still at 4.5 billion tonnes. Consumption has therefore doubled in 35 years, whereas oil consumption has only increased by half. In addition, the known reserves of coal are larger than those of oil and natural gas as they represent 120 years of extraction at the present rate.[3]

Only a short time ago, it was even estimated that world reserves would last for a further three centuries and coal was considered inexhaustible on a human scale. However, the very recent increase in the rate of extraction has changed this view, particularly as the phenomenon is likely to continue, with the most recent forecasts predicting annual extractions of eight billion tonnes in the very short term to meet the needs of the new electric power stations springing up in China and elsewhere in the world. And everything points to the fact that things will not stop there, given the ever increasing demands for electricity from many countries as they emerge from poverty one after the other.

At the current pace, the known reserves will last less than one hundred years. This leads to an entirely new, fundamental conclusion: like oil and natural gas, most of the world's reachable reserves of coal will probably be consumed during the twenty-first century, which will therefore be the century in which all the Earth's economically recoverable fossil hydrocarbon reserves are exhausted.

Given that the CO_2 emissions from coal combustion currently amount to 13 billion tonnes per year, it can be assumed that *if all the reserves are used in the same way as at present,* the corresponding emissions into the atmosphere could reach at least 2,000 billion tonnes, taking into account reserves yet to be discovered.

And indeed this is what will happen due to the frantic rate at which coal-fired power stations are being built worldwide. Even more than oil and natural gas, there is no doubt that coal will be the major cause of increased emissions, which leads to one question. Is it not possible to do something to prevent such large quantities of CO_2 being discharged into the atmosphere?

To answer the question, the first thing to note is that three-quarters of the coal used worldwide serves to produce electricity. If natural gas and oil are also taken into account, it is actually two-thirds of the world's electricity which is currently produced by fossil hydrocarbons which then release CO_2 emissions; hydraulic and nuclear power, which are emission free, produce almost all the remaining third of the world's electricity. It is therefore not surprising that electricity production from fossil fuels is responsible for nearly half (42 per cent) of the world's total energy-related CO_2 emissions and is by far the largest contributor.

There are two possible options for trying to combat such emissions, which are either to produce electricity in the future without any emissions, or to ensure that the CO_2 emitted from traditional thermal power stations during electricity production does not escape into the atmosphere. Unfortunately, the two options come up against major obstacles.

Nuclear

If we exclude hydroelectricity whose capacity is limited by geography, nuclear power appears to be the ideal solution. When Western techniques are used to harness the atom, the power stations are safe, economic and hardly emit any greenhouse gas. But the construction of nuclear plants has practically come to a standstill worldwide during the last 30 years. It is only now that some new facilities are being planned and are starting to be built in a growing number of countries.

This long period of lost interest – during which the construction of coal and gas power stations expanded without precedent – can be explained by two different causes.

The first is economic. Of course, the cost price of nuclear electricity is competitive when the prices of hydrocarbons rise. However, contrary to the forecasts, the latter were very low during the quarter of a century following the second oil crisis in 1981[4(1)]. In addition, nuclear power stations require heavy investments and take a long time to build, whereas the opposite is true of coal-fired and, even more, of gas-fired power stations. The cost overruns of more than 50 per cent on the first new generation EPR power station built in Finland by the French company Areva confirm, if this were needed, the difficulty of the task.

However, there is a second obstacle, which is the opposition of the ecologists who, in most Western countries, have convinced a large section of public opinion to oppose nuclear power station projects. Although there is currently a slight reversal in opinion, at least at the level of political leaders, we should not harbour any illusions. Nuclear power will play only a minor role in world electricity production for a long time to come, even if France has demonstrated that it is possible to produce the greater part of a nation's electricity by harnessing the atom and thus avoid the most important source of CO_2 emissions.

Opposition to nuclear power plants is incomprehensible when it comes from the very people who state that the concentration of CO_2 in the atmosphere is a catastrophe for the planet. When there are two real or alleged evils, it is vital to choose the lesser of the two, particularly as the arguments against nuclear power stations do not stand up to an even remotely serious analysis. Western nuclear power plants have proved their safety over a long period of time in countries as diverse as South Africa and China. They have never caused a single death, unlike all the other sources of energy and in particular coal, which in a country like China is directly responsible for several thousand deaths in mining accidents every year, and hundreds of thousands of others caused by the local air pollution due to coal combustion.

Moreover, the use of nuclear energy to produce electricity is recognized by international bodies as a right for all countries, whereas the manufacture of nuclear weapons is regulated by non-proliferation treaties. Over 30 countries have been operating nuclear power plants for many years, luckily without most of them having atomic weapons. Nonetheless, most ecologists do not accept this fundamental distinction between peace and conflict usage and knowingly keep alive the unreasonable fears caused by the spectres of Nagasaki and Hiroshima, as was the case with Daniel Cohn-Bendit in July 2007, when France was planning to sell nuclear power plants to Libya. According to Eurobarometer, 61 per cent of Europeans want the percentage of nuclear-generated electricity to be *reduced* for supposed security reasons.

The other arguments against the peaceful use of atomic energy do not bear close examination either. Such is the case, for instance, with the processing of by-products from nuclear power plants – derogatorily described as waste – although there are reliable techniques for processing such materials. They can be vitrified, which makes them inert for thousands of years, and then buried at depths where, whatever may be said to the contrary, they present no risk for future generations.

The same applies to the dismantling of power stations at the end of their lifespan. The operation is costly but not urgent and can be spread over decades as and when funds become available. Furthermore, the lifespan of nuclear power plants is constantly being extended. Estimated still recently at 30 years, it is now commonly set at 60 years and there is nothing to say that it could not be extended further. This is not surprising as their structure is made of heavily reinforced, extra-thick concrete to stand up to any potential incidents or accidents.

Finally, the likelihood that uranium resources will run out is not an immediate threat, as noted by the International Energy Agency in its 2006 annual report. Australia alone has identified 85 deposits, only three of which are being exploited at present, and numerous other countries also have unexploited reserves. Given the recent increase in prices, there are many projects to open more mines. In the long term, breeder reactors will push back once and for all the date when the uranium resource is exhausted.

Nonetheless, nuclear energy is one of the symbols of our time, and therefore of progress, and it is perhaps for that reason that ecologists fight against it without fear of being inconsistent. If their arguments against nuclear energy, for example regarding the treatment of waste, were to disappear, they would be sure to find others as their approach is irrational. How is it possible to assert, without batting an eyelid, that the greenhouse effect is the greatest threat ever to face humanity and at the same time oppose the main available solution, particularly as there probably will be no choice in the future but to adopt nuclear energy production on a large scale when hydrocarbons are exhausted, if necessary using breeder reactors?

As we have seen, nuclear power currently only plays a marginal role in satisfying the new demand for electricity. Of course, countries such as China have recently begun to build new nuclear power plants, but with the modest objective of increasing the share of nuclear power in national electricity production from 1 per cent to 4 per cent. Everything suggests that, by doing that, China's leaders mainly wish to stay abreast of new technology as they have split their orders between the different world plant constructors, including France's Areva, the Americans and the Russians. China is well aware that it will have to change course in the near future and start nuclear power production on a large scale. Its national coal reserves represent only 40 years of domestic consumption at the present rate, and this is increasing so fast each year that it is possible reserves will be exhausted within three decades at the most.

South Africa is another example of a country to be followed closely. Although it has very large coal reserves, it has been recently studying the possibility of constructing about ten nuclear power stations in the short term to cover 30 per cent of its electricity needs, until the project was postponed due to the 2008 fall in energy prices.

Renewable energies

Contrary to popular belief, there is little reason to be more optimistic about the potential of renewable energies for producing electricity.

Their advocates are particularly reliant on wind energy and it is true that real technical progress has been made in this field. On paper, the cost price per kilowatt-hour from wind power is sometimes competitive with other energy sources. Unfortunately, this is only in theory as wind energy suffers from a major handicap, which hampers its use. Wind turbines are by their very nature subject to the vagaries of the wind and only produce energy *approximately a quarter of the time*. And it is impossible to forecast when! The fact that they are almost constantly rotating simply gives a false illusion of efficiency. In reality, the wind speed has to exceed about 30 kilometres per hour before significant production can begin and this only reaches an optimum level when the wind speed is over 50 kilometres an hour. And if the wind is too strong, the turbines have to be stopped to prevent damage to the equipment.

Three-quarters of the time, wind turbines do not therefore produce energy and have to be backed up by traditional coal or gas-fired power stations emitting large amounts of greenhouse gases. This explains why the countries that use the most wind power are also among those with the highest CO_2 emissions.

We have also to mention the disastrous policy regarding wind turbines in France, which has no need for them at all as it already produces nearly all its electricity without greenhouse gas emissions thanks to its nuclear energy programme. A French book by Jean-Louis Butré[5] denounces the practices in this respect, as does a remarkable one in English by John Etherington, *The Wind Farm Scam*.

As for solar energy, while it can be hoped that things will change in the future, its cost price is still hugely prohibitive and it can only survive with subsidies that are even more expensive for the consumer who must eventually pick up the bill.

The corresponding cost is so extremely high for the German consumer, whose electricity bill is going to see a considerable increase – maybe of a third – because the electricity companies are obliged to buy the production by wind turbines or photovoltaic panels at prices of the order of six or seven times the cost of the electricity produced by conventional means. And so Germany is covering its territory, not only with wind turbines, which spoil its landscape, but also with photovoltaic panels. This is paradoxical in a country where there is not much wind, and that is too far north to benefit from sustained sunlight year round. But it is only the visible face of the energy policy of this country. There is another one. As the Germans gave up building nuclear power plants and as they need electricity, they build with great discretion coal-based power plants that emit a lot of CO_2.

The situation is the same in China. On 22 September 2009, the President of the Chinese Republic, Hu Jintao, announced in New York in front of the United Nations that his country's objective was that the part of the renewable energies – nuclear power and hydraulics included – reaches 15 per cent in 2020. There are no big calculations to be made to see that the electricity produced from power plants working mostly with coal will still answer 85 per cent of the needs. It is also necessary to say that one of the reasons for the interest shown by China in electricity of solar or wind origin is due to the prospect of exporting equipment to the countries where these energies are massively financed by the taxpayer through subsidies.

The same day, President Hu Jintao also announced that China was going to improve its 'energy efficiency' by about 50 per cent between 2005 and 2020, a statement which was met with very laudatory comments. But this effort, even if it is praiseworthy, must be placed in its context. Split over 15 years, it indeed corresponds to an annual improvement of less than 3 per cent a year, while the Chinese economy grows by 8 to 10 per cent a year. In other words, energy consumption and in particular that of coal will continue to progress at about 6 per cent a year and will consequently far more than double before 2020 by comparison to 2005!

Carbon Capture and Sequestration (CCS)

We thus should not harbour any illusions: most of the electricity needed for the world to eliminate poverty and to pursue its development will continue to be produced in traditional coal, gas and, more rarely, oil-fired power stations for a long time to come, hence producing massive quantities of CO_2.

This raises the question of whether it would be possible to capture the CO_2 within the power plants, transport it to a safe place, before injecting it into the ground where it would stay forever, or at least for a very long time. On paper, the solution is attractive. However, the technical and financial obstacles are such that there is little hope it will be implemented on a significant scale.

Capture techniques are still at an exploratory stage. In addition, there is no guarantee there will be enough locations with impermeable conditions near the power plants concerned to store the CO_2. At present, there are four partial CCS projects in operation in the world, each preventing around one million tonnes of CO_2 being released into the atmosphere per year. The first, the Norwegian Sleipner project, re-injects the CO_2 into an aqueous saline layer after it has been separated from the natural gas produced by a field under exploitation. The second, situated in Weyburn, North Dakota, transports the CO_2 emitted by a coal gasification power plant along a 330-kilometre gas duct and injects it into an oil field, thus enhancing the latter's production. The third, in In Salah, Algeria, re-injects part of the excess CO_2 from a natural gas field into the ground, as is the case in the Sleipner project.

The fourth CCS project was opened with great ceremony by the Swedish group, Vattenfall, on 9 September 2008 in the small German town of Spremberg. It is a small-scale coal-fired power plant, but it encompasses the complete chain of carbon capture and sequestration techniques. After extraction, the CO_2 is transported 350 kilometres to be buried in a depleted underground reservoir using a compression system. If the experiment is conclusive, the company will scale up and build two 500 megawatt power plants in 2013 in Brandenburg and in Denmark. If successful, the cost per tonne of CO_2 not released into the atmosphere is estimated at 60 euros, 85 per cent of which comes from the capture and compression operations, which are extremely energy-consuming.

Needless to say, all these projects are marginal when compared with global energy-related CO_2 emissions, which amount to 30 billion tonnes per year. It is therefore difficult to judge the technical potential of CCS at the present time.

However, there can be no doubt that CCS is heading for failure for another reason: its cost. If technical solutions are found, they are bound to be very expensive. The latest figures are most alarming. According to the International Energy Agency (IEA) (*World Energy Outlook 2007*), the typical cost of a capture and compression facility in a thermal power plant may vary between 30 and 90 US dollars per tonne of CO_2 but can be much higher, depending on the technology used, the concentration of CO_2 and the particular site. And these figures are just hypotheses, as no major power plant equipped with this type of system has yet been built, although a number of projects are in the planning stages. The cost would obviously be even higher still for retrofitting existing power stations.

Then, after capture, the CO_2 has to be transported to a sequestration site, usually several hundred kilometres away. The additional cost for a distance of 250 kilometres is estimated at 1.5 to 4 US dollars per tonne.

Finally, the gas must be compressed and injected and the site must be supervised for decades to be certain that there is no leakage, which would cancel out the advantage of the operation.

According to the IEA, the *lowest* possible total cost is presently estimated at 50 US dollars per tonne of CO_2 saved. It is only in the long term, after 2030, that it may be possible, but with no certainty whatsoever, for the cost to be reduced to 25 US dollars per tonne. Whatever the future cost may be, one thing is certain: it will be considerable. The investment required for each power plant would *more than double their cost*. That is why the 'FutureGen' pilot project in Illinois, heavily subsidized by the American Department of Energy, was abandoned when the sponsors, despite their good intentions, realized that the amount to be invested had to be increased from 830 million to 1,800 million US dollars.

At the world level, electricity production in coal, natural gas and oil-fired thermal power stations currently releases over twelve billion tonnes of CO_2 per year into the atmosphere, and this quantity is constantly increasing as they grow in number. The sums are then vertiginous. If the technical obstacles can be overcome and even if the additional cost required to prevent one tonne of CO_2 being released into the atmosphere does not exceed an average of 30 dollars, which is highly optimistic, more than 360 *billion* dollars will have to be found *every year* in order to stop CO_2 emissions and sequestrate the CO_2 produced. This is clearly out of the question.

It should also be added that the greater part of this amount would have to be spent in China, India and the other developing countries. The latter have made it clear that they have other priorities and the money required for this type of operation should come from the rich nations, which obviously cannot agree given the astronomical sums involved! It is worth remembering that every time the G8 members meet, they have to recognize that they have not kept their promises to the developing countries, and they argue over a few tens or hundreds of *millions* of dollars that cannot be found. During the July 2008 summit in Japan, African leaders noted that only a *quarter* of the sums promised by the rich countries as development aid for poor countries at the 2005 Gleneagles Summit had actually been released. Who can believe it will be any different in the years to come after the great financial crisis of 2008 and its consequences?

In addition, new commissioning of traditional thermal power stations has reached an unprecedented pace and the annual volume of CO_2 emissions can be expected to rise from 12 to around 20 billion tonnes in the near future, which will further increase the amount of money which would be needed to prevent it from being released, to well over 500 *billion dollars per year in the most optimistic scenario*.

We must therefore be realistic and not expect carbon capture and sequestration techniques to significantly change the volume of emissions caused by the use of coal, natural gas and oil for producing electricity or for any other industrial use. There is apparently only one solution for stopping emissions from coal: prohibit its extraction and close the electrical power plants that use it. Who can believe this to be possible?

Inevitable doubling of CO_2 concentrations

In conclusion, we must accept the obvious. The planet's reserves of oil, natural gas and coal are destined to be used by humankind to provide the energy needed for our development, and the CO_2 produced in the process will be discharged into the atmosphere. The combustion of all three products until their reserves are exhausted will probably release around 4,000 billion tonnes of CO_2, mostly in the next half century. To this already impressive amount should be added about 500 billion tonnes from non-energy sources, notably deforestation and farming activities. As for carbon capture and sequestration techniques, even if the technical obstacles are one day overcome, they will not be able to change the overall picture in any significant way.

These additional 4,500 billion tonnes have to be compared with the 2,800 already present in the atmosphere, although one point must be noted. The experience of past decades has demonstrated that all the CO_2 released in a given year is not to be found in the atmosphere the following year, as about half is absorbed by the natural environment and in particular by the oceans. According to the Global Carbon Project, this natural regulation has absorbed 55 per cent of the emissions in the most recent period. If nothing changes, 'only' half of the expected 4,500 billion tonnes of emissions will be found in the atmosphere, which nevertheless brings the quantity present around the Earth to a total of approximately 5,000 billion tonnes. Of course, we should not delude ourselves about the accuracy of such long-term forecasts as the figures can only be confirmed in the very distant future and are only very rough estimates.

But one thing is clear: we have to expect the stock of CO_2 floating above our heads to more or less double during this century. Expressed in 'parts per million' (ppm), the traditional unit used to show the composition of the atmosphere, the concentration of CO_2 in the atmosphere will reach around 700 ppm compared with the current level of 380 ppm.

It has to be added that although the usual forecast is for hydrocarbon fields to be exhausted in the present century, some people think there are still large deposits to be discovered, which would imply that CO_2 concentration could even exceed these levels in the future.

New techniques were very recently developed to exploit 'unconventional' oil and natural gas deposits, which were previously inaccessible. Some allow us to drill at very great depths under the oceans, and others will work horizontally to break rocks and release the natural gas they contain. These various techniques are giving us access to resources that were unknown not long ago: shale gas; oil sands; coal bed methane; heavy oils...

So, the United States, contrary to all forecasts, is today confronted with a glut of natural gas which caused a sharp fall in prices. The evolution is such that certain experts assure us that the United States have enough natural gas at their disposal for many decades to come, when they had previsouly envisaged having to equip some of their seaports with the ability to import liquefied natural gas from Africa, Russia, or the Middle East.

At the world level, the same experts assert that the new techniques will double the recoverable quantities of natural gas or oil. Of course, there is nothing certain about it, even if, past experience leads us to be careful. Did not we assert 40 years ago that oil would be exhausted 30 years later?

If these new forecasts are to be exact, it would indeed be very good news for developing countries. It would mean that mankind would have at its disposal more fossil energy and at a more affordable price than today. This would delay the moment when CO_2 emissions would begin to decline during the twenty-first century, but without questioning the general look of their evolution during the decades to come.

Given that the three fossil hydrocarbons are responsible for the great majority of greenhouse gas emissions of human origin, the latter will inevitably peak and then decline when the oil, natural gas and coal reserves are gradually used up before finally running dry.

Greenhouse gas emissions of human origin will therefore continue to grow during the first decades of the twenty-first century, but by the end of the century they will probably have more or less disappeared. Oil, natural gas and most of the coal reserves will have run dry. Deforestation will have stopped well before then. Where then would most of the emissions of CO_2 come from? The emissions are bound to peak, probably around the middle of the century, before falling. It is certain that the G8 target of halving manmade (anthropogenic) emissions will be achieved one day, but it is likely to be around 2100 and not 2050!

Needless to say, this observation is of fundamental importance. We cannot say that oil, natural gas and even coal are running short and at the same time state that greenhouse gas emissions will continue to grow indefinitely. The two issues are inextricably linked and the catastrophic scenario of endless growth in emissions is simply absurd. The atmosphere cannot receive more CO_2 of fossil origin than the total of the planet's underground carbon reserves! And, as we mentioned above, it will in fact receive even less due to the natural regulation that absorbs half the emissions.

The time has come to face reality in all its different aspects. In three centuries, the nineteenth, twentieth and twenty-first, humanity will have used nearly all the recoverable resources of fossil energy – oil, natural gas and coal – that accumulated over hundreds of millions of years. There are two consequences. The stock of CO_2 in the atmosphere, which contributes to the greenhouse effect, will more than double, rising from 2,000 billion tonnes before the industrial era to around 5,000 at the end of the twenty-first century, but it will then stop growing. The two graphs (Figures A and B) thus shed a new light on the century ahead.

Anyone interested in the fate of human beings should have no regrets. Increase in CO_2 concentration in the atmosphere is the price to be paid for enabling the greater part of the world's population to escape from poverty, destitution, disease and death in just a few generations, as did the inhabitants of today's developed countries. We have to remember that it is thanks to the use of fossil energy that we achieved living conditions that could never have been dreamt of in past centuries.

Before going any further, one question automatically springs to mind. Why is it that the data in the previous pages, which are easily understandable for anyone, cannot be found anywhere else? None of the international organisations that are supposed to guide the decisions of the political and administrative leaders – and also world opinion – clearly mention these facts which shed a new light on the debate and make it quite easy to understand.

There is a simple if partial explanation for this. Climatologists who study the composition of the atmosphere and its evolution do not express themselves in tonnes, but in concentrations. As

Evolution of CO_2 energy emissions

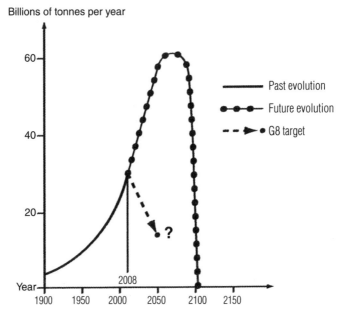

Fig. A According to the reference scenario of the International Energy Agency, annual CO_2 emissions due to the use of coal, oil, and, natural gas are going to double in the middle of the century, due to emerging countries' needs.

Everything will change later on, as oil, natural gas and even coal fields will come to be depleted and emissions disappear.

The lack of realism of the target set up by the leaders of the great developed countries (G8), which is to divide by two the amount of global CO_2 emissions in 2050 is clear. To reach it, it would be necessary, not only to spend huge amounts of money in developed countries, but to stop the development of the rest of the world and leave it in poverty.

previously mentioned, they describe the amount of CO_2 in the atmosphere using units under the mysterious acronym of ppm (parts per million), which is incomprehensible to the non specialist. Energy experts use Gt (gigatonnes) to calculate the figures for CO_2 emissions resulting from human activities – units that are just as

Presence of CO₂ in the atmosphere

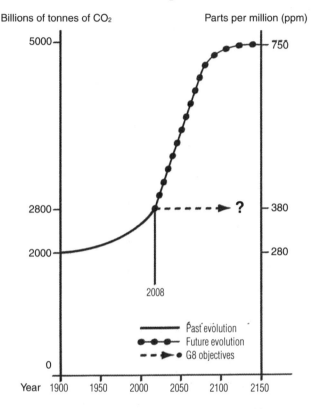

Fig. B Without the greenhouse effect, in which CO₂ undoubtably plays a small part, our planet's average temperature would be lower by perhaps as much as 30°C. Before the industrial age, the level of CO₂ in the atmosphere remained stable at around 2,000 billion tonnes, a concentration of about 280 parts per million (ppm). This concentration has increased over the past two centuries due to the burning of coal, petrol and natural gas, which has driven economic progress. Once the earth's supplies of fossil fuels have been exhausted the concentration of CO₂ in the atmosphere will stabilise before decreasing slowly over the coming centuries. The G8's objective, therefore, of maintaining the current atmospheric CO₂ concentration seems completely disconnected from reality.

mysterious for ordinary mortals. The conversion from one unit to the other is not actually very difficult as it is well established one additional ppm is equivalent to 7.3 billion CO_2 tonnes in the atmosphere. When the same thing is expressed in ppm on one hand and in Gt on the other, who is capable of following? The experts could not make it more incomprehensible if they tried. By failing to use the same unit – billion tonnes for instance – when they describe the emissions in the atmosphere and the quantities present in it, the issue is never very clear.

Perhaps it is an example of the lack of communication between different professions, in this case climatologists and energy experts, neither having made the effort required to explain things properly. However, the International Energy Agency corroborates the above conclusions as far as the evolution of emissions and concentrations is concerned. It adopted them in its *reference scenario*, before being asked by the G8 leaders to draw up hypotheses in line with the unrealistic targets for reducing global emissions that we mentioned earlier. The IEA reference scenario rightly predicts that annual global energy-related CO_2 emissions will increase from 30 billion tonnes today to about 62 billion tonnes in 2050 (*IEA work for the G8 – 2008 messages*) and that the concentration of CO_2 in the Earth's atmosphere will consequently reach roughly 600 ppm by that date.

If we remember that the targets set by the G8 in Hokkaido were, first, to halve CO_2 emissions by 2050 compared to current levels (i.e. 15 billion tonnes per year instead of 30) and second, to stabilize the atmospheric concentration of CO_2 at 380 ppm, which corresponds to today's level, we can measure the scale of the task ahead, or more precisely rather how unrealistic it is. To achieve these targets, we would have to immediately stop third world development, agree to prohibitive and unrealistic costs in developed countries, and decide not to exploit the resources accumulated in the Earth's depths over the geological eras. Who can believe for one minute that that will happen?

This sort of reasoning – which scientists describe as 'proof by contradiction' (from the Latin *reduction ad absurdum*) – is enough to reduce the targets set at the highest levels to nothing. Proof by contradiction is one of the most characteristic expressions of common sense, but everyone since Descartes knows that the latter is not very well distributed.

In dictionary definitions, proof by contradiction consists of '*disproving a proposition, not directly, but by showing that it leads to absurd or untenable conclusions*'. In the case in point, the reasoning is simple. As it is unrealistic to imagine that humanity will leave any of the world's remaining recoverable reserves of hydrocarbons unexploited, it follows that it is not possible to reduce the emissions resulting from their use. There would be only one solution for eliminating or reducing emissions, which is to close the majority of oil wells and natural gas fields and deliberately leave them unexploited, and to do the same with most coal mines, although new ones are currently being opened in Australia, Indonesia, Scotland and elsewhere. The lack of realism in this hypothesis speaks for itself.

Oil, natural gas and coal are too precious not to be fully exploited, particularly given their crucial role in helping the majority of the men, women and children living in our world to escape from poverty. And it is pointless to think about what we could do ourselves to 'save the planet' by reducing the gas emissions we produce. Our efforts would serve no purpose from this point of view: any fossil energy that we leave unused underground will be used by somebody else instead, and CO_2 emissions will remain the same at the global level.

The best we can hope for would be for emissions to be put off for a period of time, but even if that were the case, the scale of things is such that it would have no impact. For instance, Europe releases four billion tonnes of CO_2 per year, but we have seen the stock of the product in the atmosphere has already reached 2,800 billion tonnes and is growing by 15 billion tonnes per year. Even if the European Union managed to meet its official target of reducing its annual emissions by 20 per cent – i.e. 0.8 billion tonnes – by 2020, nothing at all will be changed at the global level due to the differences in magnitude, and also because most of the hydrocarbons the Europeans would not have used, would have been used by others!

Have those – including the heads of state of the richest countries in the world – who ask the world to act 'at all cost' considered for a moment what such restrictions actually would mean in practice? To satisfy them, would large sections of humanity have to continue to be deprived of electricity and stay plunged in darkness and poverty? To what level should the price of oil be raised (500 dollars a barrel or more?) in order to be consistent with their intentions? What would the consequences be in developed countries on everyday lives, and particularly on the lives of the poorest people who could never pay such prices?

Regarding people living in France, the United Kingdom or in other developed countries, why do we spend tens of billions of euros every year on 'saving the planet' since there is nothing we can do about it? In fact, French people can do far less than others, as they already have the lowest per head emissions in the leading developed countries, due to their many nuclear power plants.

Faced with such inconsistency, if not absurdity, a question emerges that deserves an answer. How has it been possible for the whole world to be misled to the point where the most elementary common sense has disappeared?

Kyoto, a senseless protocol

In 1997, no fewer than 188 countries met in Kyoto to draw up the now famous protocol. Among the signatory countries, 38 industrialized nations (so-called Annex 1 countries) committed themselves to reducing their CO_2 emissions by 5.2 per cent compared with 1990 levels, by 2008 or at the latest by 2012. The European Union even promised to reduce its emissions by 8 per cent, with France agreeing to maintain its emissions at their previous very low level, which it has indeed successfully done.

The protocol only officially entered into force in 2005, after being ratified by Russia. *But it has served no purpose whatsoever.* Admittedly, the European countries which are responsible for approximately an eighth of global emissions have more or less kept their promises. But at the world level, energy-related carbon dioxide emissions rose from 21 to 30 billion tonnes between 1990 and 2008 (up 43 per cent)!

The reason for this is that countries such as Canada and Australia and, above all, the developing countries, which did not have to sign the Kyoto Protocol, have increased their emissions massively. China's have more than doubled.

Global emissions have never grown so fast. Described by the French scientist and former minister, Claude Allègre, as being 'one of the most absurd international treaties ever to have seen the light of day', the Kyoto Protocol served no purpose and was doomed to fail. This was inevitable as it contains a basic flaw: its undertakings only concerned developed countries. In the minds of its designers, the latter would set an example to the rest of the world, which would then follow them by also reducing emissions.

Although the reasoning appeared logical, it was not. It was a sophism. Of course, after much very costly effort the developed countries could possibly control, or even reduce – slightly – their greenhouse gas emissions. But this will have no effect on the overall volume of global emissions, which will remain unchanged as the oil, natural gas and coal which the developed countries do not use will be consumed by the rest of the world. For India, China and the third world in general, it is not a question of lifestyles, but of life and death for their populations. As the Chinese expert Qin Dahe, head of the China Meteorological Administration and then Chinese representative for such affairs at the United Nations, stated: 'We lack the technology and the financial resources. Converting to cleaner energies would mean spending prohibitive sums of money we do not have. We need coal to develop. We are aware of the problem and concerned about the risks, but we have no choice' (*International Herald Tribune*, 7 February 2007).

On 9 January 2009, Agence France-Presse released the news that China has decided to increase its coal production by 30 per cent by 2015 to meet its needs, which alone will increase its CO_2 emissions by 1.6 billion tonnes per year, i.e. about twice the total reduction the whole European Union has targeted at very high costs for 2020 and will probably be unable to achieve. And this is for one just country.

India says the same thing. After the publication of the Stern report, which we shall come back to in the next chapter, an official group of economists responded, stating that: 'It was not possible for India to commit to an emissions ceiling. Its priority should be national objectives in the front line of which is eradicating poverty' (*The Indian Express*, 3 February 2007). This reaction is easy to understand, given that 400 million Indians do not have access to electricity and that every year 500,000 Indian women and children die from lung cancer caused by the smoke from the hearth in the single rooms in which they live.

For seven years, George Bush has said that the United States would not apply the Kyoto Protocol until India and China did too. He was basically right, but his explanation was wrong. He was right to refuse to commit the United States to expenditure when most of it would have been completely wasted. But instead of saying that the United States would reduce its emissions when China and India did too, he should have explained that these countries could not reduce their emissions at all and will in fact have to sharply increase them and that it was therefore simply impossible to try to fight against worldwide greenhouse gas emissions. His failure to do so has meant that the moment of truth has been postponed. It was to happen at the end of 2009, when the world community tried – in vain – to find a way to follow on from the Kyoto Protocol, this time with the participation of the emerging countries.

Copenhagen: the turning point

The objective established for the Copenhagen summit was as ambitious as it was absurd. In the application of the 'decisions' taken beforehand by heads of state at the G8 and G20 meetings, as well as the 'requirements' of the Intergovernmental Panel on Climate Change (IPCC) of the United Nations, the objective involved seeing how the planet's CO_2 emissions could be halved by 2050, beginning with the determination of an intermediary target for 2020.

However, all it needed was to add together the number of projects underway in emerging countries to realize that rather than moving towards a reduction in global emissions, the planet was actually heading for a swift, inevitable increase in emissions.

China's emissions are set to rise by more than a half by 2020. Not only are they continuing to build one or two high-powered (1,000 MW) coal-fired plants every week, but the number of cars and trucks on China's roads is set to triple by 2020. After becoming the world's leading manufacturer of lorries, cars and boats, the country is now even developing an aeronautics industry. Would it be doing this only to then leave its power stations unused, its vehicles in their garages and its aircraft on the ground?

Considering that China is building 5,000 kilometres of motorway each year, that Beijing is completing work on its seventh motorway bypass and that only 300 million people in China, out of a population of 1.4 billion, are beginning to acquire a Western standard of living, one can see why a sharp increase in China's emissions is inevitable.

The four billion inhabitants of India, Brazil, Indonesia and the rest of the emerging world are moving along the same road. Ongoing projects in these countries mean that their emissions will also increase at a brisk pace over the coming decades; it is the condition for lifting their populations out of poverty. 400 million Indians still have no access to electricity. So how can they cut emissions they do not produce? India, like China, has vast coal deposits that are readily accessible due to an abundant, low-cost workforce. We need to look reality in the face. For these countries it is the only way to produce the affordably-priced electricity that they so desperately need.

Notwithstanding the speeches of Barack Obama, the United States can reduce their emissions only marginally (4 per cent in 2020 compared with 1990). Homes there are twice as large as in Europe, travel distances are twice as long and coal is needed to produce half their electricity. And as we all know, they have a continental climate. Without air-conditioning, the southern half of the United States would be uninhabitable. In contrast, Chicago can experience winter cold spells with temperatures below -30°C. It should therefore come as no surprise that energy-saving opportunities on the other side of the Atlantic are limited, although not completely nonexistent, when it comes to road vehicles for instance.

Finally Europe, with an eighth of global emissions, can actually do nothing. Even if it could eliminate its emissions entirely, the impact would be minimal at the global level and quickly offset elsewhere.

So the problem was unsolvable and the heads of state and government gathered in Copenhagen provided a dumbfounded, incredulous planet with a ludicrous performance that attempted to square the circle. The Europeans were undoubtedly the worst at this game. By affirming that curbing emissions was simply a matter of 'political will' rather than the physics (if not elementary mathematical) problem that it actually is, they demonstrated a flabbergasting level of ineptitude. How could they hope to divide something by two when all the signs pointed to its being multiplied by two?

Contrary to what was widely reported, on this occasion it was the Chinese president, Hu Jintao, who rendered the greatest service to humanity. From the outset, he was the only head of state who refused to go to Copenhagen to take part in what he knew would be a farce. In his place, he sent his prime minister, Wen Jiabao, with a dual mission. Firstly, he was to refuse any detailed commitments to reducing China's emissions. But his mission also included refusing any global emissions targets that would obviously have had a direct impact on China, the country with the world's highest emissions.

Wen Jiabao carried off his dual mission with the utmost scrupulousness, with the fundamental consequence that Copenhagen ended in total failure. Nothing now remains of the absurd edifice that had been taking shape over 20 years and that wanted the planet to reduce its emissions whilst continuing to develop totally contradictory objectives.

It should be pointed out that China was not the only country to scupper the Copenhagen summit. It had the support of India and nearly all the emerging countries that did not want a curb on their development. It also had the tacit support of President Obama, who was only too glad not to be held responsible for a failure that he knew then to be inevitable. The United States Senate would never have accepted the restrictive measures and the exorbitant financial cost of reducing significantly the country's emissions whilst the rest of the world increased theirs.

And so, at the beginning of 2010, the 'Cap and Trade' project Obama announced at the beginning of his mandate seems sure to fail. The principle of the operation is attractive at first sight. The idea is to set a ceiling to the overall CO_2 emissions of the industrial sector, and to reduce it year on year. In the frame of this overall ceiling, each industrial company would be given a quota of emissions not to exceed (cap). Should the company use more than their given quota, they could buy (trade) 'emission rights' corresponding to the extent to which they have exceeded the ceiling, to other emitters who have not reached their own quota. Such a device exists in Europe, where a 'CO_2 market' has so been created.

But, if the principle seems simple, the application of the project is very difficult, because everything depends on the overall ceiling. If this is too high, the whole project is of no use, and the value of a tonne of CO_2 on the market of 'emission rights' collapses. This is what took place in Europe in 2007. Equally, if the ceiling is too low the opposite occurs. The value of a tonne of CO_2 can swiftly rise, and the companies that cannot keep within their quota are very heavily penalized.

To be effective, the 'Cap and Trade' programs should give priority to the main emitters, that is the coal-fired power plants which produce half of the electricity in the United States, and give them a reduced quota. The consequence would be immediate: the producers of electricity would spend considerable sums buying emission rights to reduce the extent to which they have exceeded their quotas.

In other words, the price of the electricity would increase greatly. It is easy to understand why the elected representatives of the numerous American states that possess coalmines are the most convinced opponents of 'Cap and Trade', because they know there is no technical means of preventing the combustion of coal from producing CO_2. These price increases would also concern heavy industry, such as cement factories. For many of the largest emitters of CO_2, this project will simply appear as an additional tax.

The opposition of the concerned representatives is all the more violent as they know that these heavy industries need to compete with industry in Asian countries who will not be introducing such a scheme.

Furthermore, 'Cap and Trade' is doomed to fail in the US due to the adoption, by the US Senate in 1997, of a resolution that stipulates: 'The United States shall not sign any protocol which would lead to limit or reduce greenhouse gases emissions in developed countries without similar requirements being accepted by developing countries, the absence of which would result in significant negative impacts for the American economy.'

It was this resolution that prevented President Clinton and Al Gore from adopting the Kyoto Protocol but it applies equally well to the 'Cap and Trade' project.

In the end, it is only Europe that does not understand the complexities of the situation. Governed by politicians for whom ideas take precedence over facts, it continues to want to 'set an example' for the rest of the world, even though it is merely castigating itself and dramatically hampering its economy in the global competitive marketplace under the illusion of 'saving the planet'.

Should we all become vegetarians?

Will we all have to stop eating meat one day? The question may seem ludicrous, but some say that eating meat may seriously damage the environment for two reasons that are added together.

First, it takes from five to ten kilos of cereals to produce just one kilo of meat, which is a very poor use of the planet's resources. Second, ruminants may have another impact on the environment, as noted in the report by the United Nations Food and Agriculture Organisation (FAO), in November 2006, which was widely covered in the news at the time. The report revealed, 'Animal farming produces more greenhouse gases than all means of transport together, with 18 per cent, measured in CO_2 equivalent, compared to 16 per cent of global emissions.' Moreover, present trends suggest that the world production of meat will double by 2050, giving rise to seemingly worrying forecasts.

Cattle are mainly responsible for this situation, as they release large quantities of methane, a gas that is said to be 23 times as warming as CO_2. In France, as surprising as it may seem, a cow releases more greenhouse gas every year on average than a car. More precisely, according to the United Nations experts' calculations, a cow produces 106 kilos of methane when chewing the cud, plus some CO_2 and nitrous oxide, totalling an equivalent of 2.5 tonnes of CO_2. An average car in France travels 14,000 kilometres per year, during which it releases 165 grams of CO_2 per kilometre, i.e. 2.3 tonnes overall, a figure that has been falling regularly year after year.

As there are 21 million cattle and 30 million cars in France, a question naturally springs to mind. If we absolutely must reduce our greenhouse gas emissions as we have been told, would it be easier for us to give up our cars or to stop eating beef?

The answer is obvious since cars are now a vital part of everyday life. Questioned by the Institut Ipsos in January 2007, the majority of French people (55 per cent) quite rightly considered that cars were more essential to their lives than steak, although it is their national dish! To be coherent, ecologists should give priority to asking their fellow citizens to give up eating beef, which they can do without, and not to depriving them of their cars, which are nearly always essential to their everyday lives. But, with well known exceptions, the ecologists have doubtless realized that this would not be very popular.

If we all became vegetarians, the contribution of agriculture to the greenhouse effect may perhaps fall. However, as this hypothesis is hardly credible, we must expect an increase in the corresponding atmospheric emissions with the decline of world hunger and the development of animal farming which, except for vegetarians, always goes hand in hand with higher standards of living. These prospects are all the more probable since the current world trend shows an increase in the average weight of cattle and consequently their unit emissions, contrary to what is happening with cars for which average consumption is falling! In short, trying to reduce emissions caused by animal farming does not appear for the time being to be any more realistic than attacking those caused by the use of hydrocarbons.

The Gerondeau Paradox

In the end, we need to call into question our previous opinions on greenhouse gas emissions as we are confronted with a fundamental revolution in our ideas. After reading the previous pages, a number of questions have to be asked: In terms of reducing CO_2 emissions – almost an obsession at the beginning of this century – what is the use of:

- insulating housing
- taxing fuel used for land, air and marine transport
- investing in renewable energies
- setting up a permit trading system
- taxing road transport
- investing in railways
- trying to discourage people from using their cars
- making the use of low consumption light bulbs compulsory, etc.

- reducing investment in roads
- trying modal shift actions
- using bikes
- providing heavy subvention for bio-fuels, wind farms, solar panels
- reducing fuel consumption standards on cars
- editing international protocols
- issuing European Union's drastic targets for reducing CO_2 emissions
- launching 'major works' programmes
- enforcing national, regional or local 'climate plans'
- inducing individual efforts to reduce emissions, etc.

What is the point of all these measures from the point of view of tackling global CO_2 emissions? NOTHING.

Nothing from the point of view of global greenhouse gas emissions and therefore of any possible impact on the climate: absolutely nothing, as quantities not emitted by some people or some nations will be emitted by others. This is a Copernican revolution in terms of our perception of one of the major planetary phenomena of our time. We have been reasoning as if the world had an infinite stock of hydrocarbons, which may perhaps have justified trying to limit our emissions by leaving a sizable part underground forever. As that is not the case and reserves are limited and will in any case be fully exploited, our efforts no longer make any sense. At best, they would put off some emissions for a few months or a few years, which would not have any sizable effect in terms of CO_2 concentration, in view of the fact that there is such a considerable stock in the atmosphere (close to 3000 billion tonnes vs. a monthly increase in the range of one billion). Even if there is a time lag of ten years or more, which is impossible, this would make no significant difference whatsoever.

Of course, this does not mean that some of the actions listed above are not justified, but this is for reasons other than wanting to fight CO_2 emissions. It can be a good investment to buy cars that consume less fuel, or to insulate certain parts of houses. For governments, it is sensible to prepare for the post-oil period and it is therefore reasonable to invest in the appropriate research. However, we should not have any illusions: only a small minority of the endless list of expenditures today supposed to 'save the planet' is justified; it is important not to confuse two unrelated issues: the need to limit our energy expenditure and prepare for the future, and the impact of our actions on CO_2 emissions.

Of course it is right to make savings to reduce our expenditure and not waste global resources; and it is right to prepare for a post oil world in the second part of this century. But to imagine that the CO_2 that may be saved in the process would contribute to 'saving the planet' is science fiction, as it will be emitted by someone else, in China or elsewhere. Squandering taxpayers' or consumers' money on the basis of this illusion is unjustified.

Earth's carbon footprint

There is something deeply paradoxical about this. *It is not easy to understand that when you reduce your CO_2 emissions, you are in fact reducing nothing at all.* To quote the famous saying of the nineteenth-century French economist Frédéric Bastiat, there is 'what is seen and what is not seen'. We see our own emissions, and they are indeed falling. What we fail to see is that other people's emissions increase by as much, so that the total remains the same. It is a constant sum game as all the planet's fossil hydrocarbons will be used and the carbon they contain will produce CO_2. The 'global carbon footprint' of the planet cannot be changed, as it is a direct and unavoidable consequence of the amount of carbon left underground in ancient eras.

René Descartes was maybe right when he wrote in *Discourse on the Method*: 'Nonetheless, a plurality of voices is not a proof worth anything for truths which are a little difficult to discover, because it is far more probable that one man alone himself would have found them than an entire people.'

Curiously, until now we have studied the consequences, i.e. CO_2 emissions, and not the causes, i.e. the extraction of oil, natural gas and coal. Governments worldwide are even encouraging efforts to discover and exploit new reserves, which obviously dooms all their ideas of reducing global emissions to failure.

All the efforts currently being attempted – in vain – to fight CO_2 emissions are also often presented as the present generation's contribution to improving the lot of future generations, a 'solidarity pact' between them. Unfortunately, it has the opposite effect. Pointlessly spending money can only slow the progress of the economy and have a negative impact on the standard of living of our successors and on poverty reduction worldwide.

Thinking about it carefully, the fact that it is impossible to fight against CO_2 emissions may have one positive side, which is to help avoid considerable expenditure for individuals, companies and the public finances. In line with other European countries, France alone has already granted or planned absolutely gigantic amounts of public expenditure designed to contribute to an illusory world effort to reduce CO_2 emissions. Subsidies for wind turbines, solar panels, bio-fuels, rail and public transport, large scale insulation for every building etc. can be counted in *tens of billions of euros per year* and will increase almost endlessly if the *Grenelle de l'environnement* (Environment Summit) decisions are implemented. Further tens of billions of euros are also being paid, or will be in the future, by individuals and companies who will face higher bills and taxes, and be compelled to keep increasingly numerous, stricter standards, on the pretext of saving the planet – and to the benefit of vested interests. And the situation is the same in other countries.

In the light of the observations made in this book, massive savings are possible for the public finances as well as for private individuals and companies. Instead of being wasted in the misconceived belief that it can have an impact on worldwide CO_2 emissions, the money could then be used to increase our fellow citizens' spending power instead of reducing it, or to *really* help the world's poorest populations, including those said to be under threat from climate change. As we cannot influence the latter, we could at least try to limit its possible consequences.

In this major issue, one of the leading causes for concern at the beginning of this century, we have to understand why the whole world has gone off track. But, firstly, we must also answer another nagging question. As we can do nothing about global CO_2 emissions and as its concentration in the atmosphere is inevitably going to double, are we heading for disaster?

***

Before moving on to the next chapter, there is one last remark. People who contest the ideas *en vogue* on the impact of CO_2 emissions on climate change have had, until recently, the utmost difficulty making themselves heard. They come up against organizations and people who control public opinion because they are invested with official labels (United Nations, governments) which give them credibility. Seen from the outside, any challenges to the official line on climate change are necessarily seen as just battles between experts, and it is the ones with the label of the authority, and with the corresponding fundings, who win and have been, until now, trusted by the media and the public.

On the other hand, the same 'official experts' have always said that we will soon run out of oil and have managed to convince the public of this. It would probably therefore be wise to change battleground and attack the debate from a different angle, by asking the question: 'Can we reduce global CO_2 emissions?' and by putting forward the evidence, i.e. that the only realistic way of reducing emissions would be to leave part of the world's underground resources of oil unexploited, which is impossible for anyone to believe, the same being true for natural gas and coal.

A different set of specialists should then take part in the debate: geologists, oil experts, energeticists, economists. When it becomes clear that there is nothing we can actually do about global CO_2 emissions, and that CO_2 concentrations in the atmosphere will necessarily almost double in the present century, it is surely easier to get across the message that changes in levels of CO_2 do not necessarily have a negative impact on the climate. Nobody likes the idea that we are heading for the end of the world. As for the usual claims that 'we have ten years left to act', they will automatically disappear once it is recognized that we can do nothing to decrease emission levels. Finally, the emerging countries, starting with China and India, would probably back this new approach.

Maybe the 'sceptics' are fighting the battle on the wrong ground, that of their opponents, and it would perhaps be a good idea to change to another.

At this point of the book, we cannot fail to pose another question. Is it quite necessary to go any further? Does the debate, which has on the one hand those who are convinced we are living in a time of global warming caused by mankind and on the other those who are sceptical of this theory, have any utility since we cannot do anything significant about global emissions? Isn't it something akin to the controversy of the sex of the angels? Will future generations not see it thus?

We must go on because the consequences of the 'official' theses on climate change are so grave, not just for the future of the global economy but also for the vision we have of the world in which we live.

2   Should we trust the IPCC?

One of the keys to understanding the issues presented here is the role of the different players in this global human comedy: climatologists, economists, energy specialists, ecologists, politicians from developed and developing countries, official international agencies, a host of pressure groups, not forgetting all sorts of headline-hitting alarmists.

The IPCC

In 1988, the World Meteorological Organization (WMO) and the United Nations Environment Programme (UNEP) chaired by Maurice Strong made the joint decision to set up a new body, the Intergovernmental Panel on Climate Change (IPCC). It was given the mission of gathering scientific, technical and socio-economic information relevant to understanding climate change due to mankind activities and its potential impacts, and to identifying possible options for adaptation and mitigation.

The IPCC set up three working groups, dealing respectively with understanding the phenomenon, the options for adapting to it and the means of mitigating it, as well as another body designed to develop appropriate tools for measuring each country's greenhouse gas emissions. The members of these different groups are appointed by national governments and by various scientific organizations and environmental non-governmental organizations (NGOs). An elected Bureau oversees all the work, backed up by a permanent secretariat and technical support units from the different working groups.

The IPCC holds a plenary session every year to decide on its work programme and to approve completed reports. However, one of the IPCC's main activities, and certainly the one that has the most media coverage, is to publish regular 'Assessment Reports' on the current state of knowledge on climate change. The most recent documents published by the three working groups were dated November 2007 and served as a basis for the Fourth IPCC Assessment Report (AR4) also dated 2007. A 'Synthesis Report', grouping the three working groups' 'Summary for Policymakers' together, was published at the same time. This fundamental document, which is also sometimes called the 'Summary for Policymakers', can be consulted freely on the Internet, as is the case for most of the organization's reports. It should be said that this is the only document designed for reading by non-specialists.

As it is easy to imagine from this description, the IPCC's mechanisms tend to be very cumbersome. The working groups' reports are several thousands of pages long and thousands of experts contribute to them. The conclusions are adopted by a voting system, which has led a certain number of scientists to stop taking part, on the grounds that they were unable to make themselves heard.

The result is *la pensée unique*, an official way of thinking. This is quite clear in the Synthesis Report mentioned above. This document, *Climate Change 2007: Synthesis Report*, is divided into several parts. The first describes current knowledge of past climate changes and proposes projections for the future. The remainder looks at the possible consequences for the planet and for humanity.

Climate change: the past

According to the members of IPCC: 'Warming of the climate system is unequivocal, as is now evident from observations of increases in global average air and ocean temperatures, widespread melting of snow and ice and rising global average sea level.'

Their verdict seems to be final, but the alarmist tone does not really match the figures that go with it. For example, they indicate that: 'Global average sea level rose at an average rate of 1.8 mm per year over 1961 to 2003 and at an average rate of about 3.1 mm per year from 1993 to 2003. Whether this faster rate for 1993 to 2003 reflects decadal variation or an increase in the longer term trend is unclear.' *At the most*, the sea level is therefore currently rising at a rate of 3.1 centimetres per decade, which can hardly be termed as catastrophic! At a time when many people are telling us that the ice sheet is breaking up, the rate of sea level rise has actually slowed down to 2.3 millimetres per year since 2003.

As for the average global temperature, the document indicates that it rose by 0.74°C in the past hundred years. Once again, however things are presented, less than one degree in one hundred years is very far removed from the usual, widespread views on climate change.

Later on, the same report indicates that, from 1900 to 2005, the area affected by drought has *'likely'* increased since the 1970s.

The use of the adverb 'likely' may seem surprising, but it stems from the IPCC's proceedings. Members are asked to choose terms to express likelihood, e.g. 'very unlikely', 'more likely than not', 'likely', 'very likely', etc. and, depending on the case, to express their judgement on the correctness of the submitted data following a scale from 'very high confidence' to 'very low confidence'.

This is contrary to the requirements of a scientific approach, which involves confronting the different points of view until the truth can be found. If they had used the IPCC's methods, contemporaries of Galileo would have voted that it was 'extremely unlikely' that the Earth goes round the Sun; those of Pasteur that the spontaneous generation of microbes was 'very likely' and those of Einstein that the theory of relativity was 'very unlikely'! There is something shocking about this because the public and political leaders put their faith in what they believe to be a scientific approach when in fact it is not[6].

The IPCC also describes the increase in greenhouse gas emissions referred to in the previous chapter and confirms that their energy-related component has risen from 15 billion tonnes of CO_2 in 1970 to 30 billion today. Comparing the rise in temperatures observed in the last two decades of the twentieth

Changes in atmospheric CO_2 concentrations

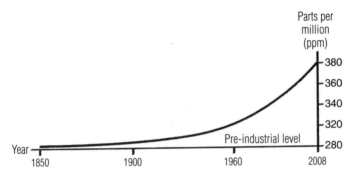

Source: IPCC

Fig. C Information is available on the gradual rise in atmospheric CO_2 concentrations, from 1960 thanks to the Mauna Loa Observatory in Hawaii.

The changes seem to be characterized by their very high regularity.

century and the increase in atmospheric CO_2 concentrations, the IPCC deduces in its Synthesis Report that: 'Most of the observed increase in global average temperatures since the mid-20th century is *very likely* due to the observed increase in anthropogenic greenhouse gas concentrations.'

Organized deceit

We must stress this crucial point, which is at the very heart of the matter. The IPCC does *not* state that the relationship between atmospheric greenhouse gas concentrations and the rise in global average temperatures is 'certain', contrary to what is repeated everywhere on its initiative, but only that it is 'very likely'.

It is easy to understand why some IPCC members had doubts and refused to vote for the term 'certain'! In reality, the relationship is 'very unlikely': when the curves for the two phenomena are compared, the results are not at all what one might expect (Figures C and D).

Changes in global average temperatures

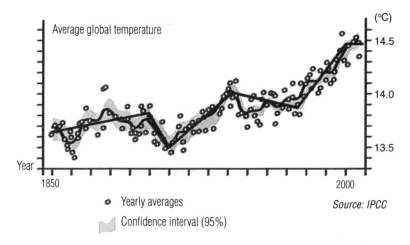

Fig. D Even at a glance, it is clear that the curve for global average temperatures since 1850 has no relationship with that for CO_2 concentrations. The curve is not regular but chaotic, with a succession of periods of stability, reductions and increases. There is no explanation, for example, for the period of cooling observed from 1945 to 1978.

Whatever the case may be, the variations are very low, with less than one degree in over one hundred and fifty years, which is far less than in other periods in the past.

We have precise figures for changes in atmospheric CO_2 concentrations. There was an extremely regular change over the past decades, from 280 ppm before the industrial era to 320 ppm in 1960, then to 380 ppm at present. In contrast, changes in global average temperatures since 1850 show a chaotic path, with a slight increase from 1850-95; a drop from 1895 to 1910; a strong rise from 1910-40; a slight fall from 1945-75; a further rise from 1975-98 and stability, if not a slight decrease, since then, contrary to generally accepted ideas (Figure E).

In other words, it is hard to imagine two more different curves than the one tracing atmospheric CO_2 concentrations and the one for temperature fluctuations.

Global average temperature change in past 10 years
Average over 150 years

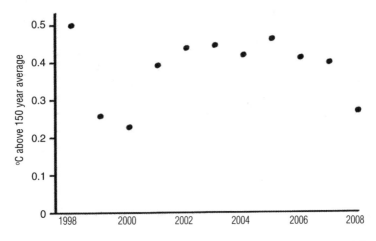

Source: Hadley Centre/Meteorological Office

Fig. E After a period of warming lasting about 20 years, from 1975 to 1998, the global average temperature has stopped rising in the last decade. This is confirmed by the various internet sites dealing with the subject, including the British Met Office whose figures are shown above and which serves as a reference for the IPCC. The temperature is currently stable at around 0.4°C above the average recorded for the last one hundred and fifty years, and is maybe tending to decrease. At a time when CO_2 concentrations have never been so high, this trend completely contradicts generally accepted ideas. This is a real case of an 'Inconvenient Truth'.

In particular, no one has been capable of explaining the *decrease in temperatures recorded from 1945 to 1975* at a time when *CO_2 concentrations were rapidly growing.* To remind you, at this time most climate experts were predicting the advent of a new cold era!

In these circumstances, how can the IPCC say that it is *'very likely'* that the increase in greenhouse gas concentrations is responsible for climate change since the mid-twentieth century, when the temperatures *decreased* for the first 25 years of the second half of the century?

Nobody has been capable either of explaining why temperatures have not continued to rise since 1998, although CO_2 concentrations have never increased more than at the present time.

It is also revealing to see that the IPCC only looks at the second half of the twentieth century, ignoring what happened before then. If the rise in CO_2 concentrations were really responsible for climate change as the IPCC claims, how can we explain the significant temperature fluctuations recorded in the first half of the twentieth century, given that the quantity of CO_2 in the atmosphere was practically stable at that time? It is obviously no coincidence that the IPCC takes 1950 as a starting point for its 'deductions' and never mentions what happened before then. If it had merely chosen 1940 instead of 1950, its entire reasoning would have been destroyed, as temperatures were higher in the 40s than in the 50s.

By misrepresenting the facts, the United Nations' representatives have managed to convince world opinion and the policymakers, including the leaders of the world's most powerful nations, of a flagrant untruth – the certainty of a link between growth in greenhouse gas concentrations and climate change – and its implicit corollary that humans are responsible for climate change, are therefore guilty and must change their lifestyles.

And yet, everybody knows that there have been many episodes of climate change in the past and that human activities had nothing to do with them. Otherwise, how can we explain that the Vikings farmed in Greenland for three centuries, proof of a warming of the climate that was very probably greater than the current one? How can we explain that Europe experienced a Little Ice Age from about 1600 to 1850, when it was possible to cross the Seine or the Thames on the ice almost every winter?

We often hear people say that a rise of less than one degree is something considerable. It is therefore all the more surprising to read what Jean Jouzel, a top French climatologist, had to say: 'Greenland ice samples helped discover 25 extremely rapid and important climate changes during the last Ice Age and in the period of deglaciation which followed it, with increases of up to 16 degrees celsius in just a few decades, followed by slower cooling'[7]. It is obviously difficult to deduce from this observation that we are currently experiencing a marked episode of swift warming. *On the contrary, everything seems to indicate that we are in a period of stable Earth temperatures, as variations are being recorded in tenths of a degree.*

More surprising still is a communiqué from the CNRS (French National Centre of Scientific Research) dated 19 June 2008. In 2007, an international team carrying out research on ice sheets in Greenland used new techniques that showed that 'the shift from one climate episode to another can occur in a single year. In this way, 14,700 years ago, there was an episode of warming with a rise in air temperature of over 10°C.'

It is therefore quite understandable that some people, like Claude Allègre, a well-known French scientist and a former government minister, believe that the IPCC has lost all legitimacy and has been taken over by proponents of *la pensée unique*, who favour the official line on ecology and no longer use a scientific approach. *Atmospheric CO_2 concentrations have increased by 40 per cent since the beginning of the industrial era. If they had a major impact on the rise in temperatures as the IPCC claims, the climate should have changed drastically, but this is fortunately not the case.* The truth is that there have always been climate changes; that those which we are currently experiencing are more moderate than in past episodes; and that there is no proof that they are induced by humans. Moreover, the atmospheric CO_2 concentrations are extremely low, as 380 ppm only represents 0.038 per cent of the atmosphere's total composition. There are many other components that could play a far greater part, starting with water vapour and clouds which, according to Claude Allègre and other specialists, have 40 times more impact than the variations attributed to the other greenhouse gases![8]

Admittedly, recent publications from NASA say that: 'the south polar ice cap has receded in recent years, evidence of certain

The south polar ice cap on Mars

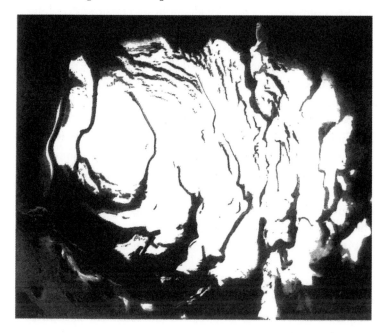

Photo Credit: NASA/JJPL-Caltech

Fig. F When American and European satellites orbited the planet Mars they found that the ice caps are changing over time. The south polar ice cap of Mars has receded in recent years, evidence, according to Nasa, of 'certain climate change' which can obviously not be blamed on human activity of any sort!

climate change.' However, those who think that this confirms the harmful impact of human activity should not rejoice too quickly, as the quotation refers to the planet Mars. It just proves, were this necessary, that there are always natural fluctuations in the different planets present in the solar system[9] (Figure F).

It is hard not to ask an awkward question at this point. Given the fact that it has received the world's highest distinction, the Nobel Peace Prize, is it possible that a United Nations body is telling falsehoods? Do we have the right to question it and form our own opinions?

I have to confess that for a long time I have been answering these two questions in the negative, as have many other people. It just seemed inconceivable to me that an official United Nations body, holder of one of the highest world distinctions, can be questioned. In a previous book, I thus made the assertion that it was 'very likely' that human activities, and more precisely CO_2 emissions, were changing the climate, and that the planet was indeed warming. Being a Frenchman, I would have done well to listen to René Descartes. The first precept in his Discourse on the Method was 'to accept nothing as true which we do not clearly recognize to be so.' In other words, we should not believe anything until we have checked it. In this century as in past centuries, repetition of the same thing by many people does not necessarily mean it is true.

In this case, there is no need to be a scientific expert to see that the curves for atmospheric CO_2 concentrations and for global temperatures are in no way related and that, contrary to the incessantly repeated claim, it is therefore unlikely that there is a significant link between the two phenomena.

As all the United Nations' forecasts are based on the hypothesis that climate change is directly linked to the rate of CO_2 in the atmosphere, it is possible that their conclusions could be proven to be false.

But even if this assumption proved to be true one day, the IPCC's works show that the consequences of climate change would not be nearly so catastrophic as often imagined.

The official forecasts

Climate change: the future

The IPCC uses hugely powerful computers to forecast future climate change that produce hundreds of models based on many thousands of different datasets. It is therefore not surprising that, depending on the initial assumptions and on the models used, the results are characterized by very wide dispersion and great uncertainties. In the last Synthesis Report, the IPCC retains six scenarios (among hundreds of others) for the scale of human-induced greenhouse gas emissions at the end of the twenty-first century. The figures range from 20 to 130 billion tonnes of CO_2-equivalent emissions per year (compared with the current level of 50 billion tonnes). And, as for the possible temperature rise at the end of this century, the dispersion is from one to six degrees celsius, which gives an idea of the models, uncertainty, not to say contradictions!

Even if we go along with the IPCC's claim that the climate depends on atmospheric CO_2 concentrations, we can be somewhat quizzical about the very high dispersion found in these scenarios, particularly the highest estimates. As we have seen, not only oil, but also natural gas and even coal reserves, would most probably be exhausted by the end of the twenty-first century. So where will the estimated annual emissions of 130 billion tonnes of greenhouse gases come from? For these figures to be right, extractions of these products would have to almost triple in volume compared with their current levels, although there will be practically no reserves left in the ground. The climatologists who approved these unrealistic hypotheses had obviously not consulted the experts in the energy sector. Or perhaps they are well aware of the physical impossibility of their claims, but are painting a black picture of the situation on purpose, so as to back up the idea that we are inevitably heading for a disaster. Or it may be a mixture of the two explanations.

Whatever the case may be, the most extreme hypotheses must be put to one side, which limits possible emissions at the end of the twenty-first century to a range of between 25 billion tonnes – the most plausible estimate given that fossil hydrocarbons are expected to have run out by the end of this century – to 65 billion tonnes of CO_2-equivalent. It is then interesting to see what the IPCC thinks the impact of the corresponding emissions will be, as they give predictions for sea level rise and increases in global temperature for each of the different scenarios.

Table 1

Sea level rise and global average warming at the end of the twenty-first century based on different scenarios (from IPCC 2007 Synthesis Report).

Scenario	Emissions in 2100 (1)	Sea level rise (2)	Temperature change (3)	
			Best estimate	Likely range
Scenario B1	25	0.18-0.38	1.8	1.1-2.9
Scenario A1 T	30	0.20-0.45	2.4	1.4-3.8
Scenario B2	70	0.20-0.43	2.4	1.4-3.8
Scenario A1B	65	0.21-0.48	2.8	1.7-4.4

(1)   Billions of tonnes of CO_2-equivalent
(2)   Metres in 2099 compared with 1999.
(3)   Degrees celsius in 2099 compared with 1999.

Rise in sea level

The estimates for sea level rise range from 18 to 48 cm, the majority amounting to about 30 cms for the coming century.

The IPCC specified that it had excluded the hypothesis of extremely rapid changes leading to much higher sea rises particularly due to a sudden but very unlikely melting of the Greenland and Antarctic ice caps.

Rise in temperatures

The same study shows that the best estimate of the 'most likely' average rise in temperatures at the end of the century is in the region of 2 to 3°C, but once again with a very wide range of possible dispersion around these values.

These are the IPCC's forecasts. Once again, it is important to remember that these are far from certain as nothing proves the relationship between CO_2 concentration and the climate. But, if they prove to be true, we still have to imagine what impact this will have on the planet and for human beings. It is hard to believe that it will be anything like the situation that is usually predicted.

So the IPCC's own work led to two main conclusions: during the twenty-first century, sea levels would rise by about 30 cm and the global average temperature would increase by 2 to 3 degrees celsius. In order to make policymakers aware of the situation, the Synthesis Report gives 32 examples of the possible impact of these two phenomena. Almost all these 32 examples are negative and apocalyptic. And yet it is hard to believe that if these changes do take place there would be no positive consequences anywhere in the whole world. There are some cold countries that would not complain if the temperature rose by a few degrees and some regions in the world that would be happy to receive more or less water depending on the case! How can everything be negative?

But other points are even more revealing.

Rising sea level and its alleged impact

The alleged impacts of the predicted 30 cm rise in sea level over the present century, i.e. 3 mm per year, are just incredible. We quote them here so that everyone can form their own opinion of the 'politically correct' stance adopted by the IPCC:

Africa: Towards the end of the 21st century, projected sea level rise will affect low-lying coastal areas with large populations. The cost of adaptation could amount to at least five to ten per cent of Gross Domestic Product (GDP).

Asia: Coastal areas, especially heavily populated mega-delta regions in South, East and South-East Asia, will be at greatest risk due to increased flooding from the sea.

Australia: By 2050, ongoing coastal development is projected to exacerbate the risks from sea level rise and increases in the severity and frequency of storms and coastal flooding.

Europe: Negative impacts will include increased risk of inland flash floods, more frequent coastal flooding and increased erosion (due to storminess and sea level rise).

North America: Coastal communities and habitats will be increasingly stressed by climate change impacts.

Small islands: Sea level rise is expected to exacerbate inundation, storm surge, erosion and other coastal hazards.

If we remember that this litany of catastrophes corresponds to a predicted rise in sea level of 30 cm in a century, i.e. 3 cm every ten years, the immediate response is to ask: who are they trying to kid? According to information published by the CNRS on the Internet, a one metre sea rise makes the coastline recede on average by one hundred metres. For three centimetres, the coastline would therefore usually recede by only three metres. And that would be of course happening only in flat areas and not on rocky seashores or in already built-up locations.

On every sea and ocean of the world, there are from time to time storms with waves five metres (about 15 feet) high or more. What difference does it make if they are three centimetres higher? Existing land is necessarily several metres higher than the sea, and a few centimetres or decimetres more or less do not matter. It's only if the sea was to rise by a number of metres (as in the Al Gore film) that we would have to be concerned, and the IPCC itself considers this impossible. So why create panic without any reason?

How can anyone seriously put forward the idea that the cost of such a small rise would be 'five to ten per cent of GDP' without specifying, for that matter, whether this was for the entire African continent, for a specific country or for a region? Where do the figures come from? Do they really know how much it would cost to construct a 30-centimetre high sea wall in a century's time, should it be necessary? Have they forgotten that a large part of the Netherlands is several metres below sea level? Or that many countries have tidal amplitudes of several metres, sometimes as much as ten? Or that even now the continents often rise or sink by a few decimetres?

Have they also forgotten that the entrance to the Cosquer Cave, discovered a few years ago in France's Côte d'Azur, is now located 37 metres under the Mediterranean Sea, whereas it used to be above sea level? 12,000 years ago, the sea level rose by 120 metres in a few centuries when the ice sheets that covered over half of Europe, North Asia and North America melted. They are not going to melt twice! Who can seriously believe for one minute that a possible rise in sea level of three centimetres per decade will lead to the series of catastrophes the IPCC describe? And even if this rise did take place, who can believe that mankind will not take the necessary measures to cope with the situation by 2100, by adding to or constructing new sea walls, given the small, not to say derisory, rise they are talking about?

Only one conclusion can be drawn from reading these predictions, and that is that the IPCC has lost the neutral position to be expected of it. It can be considered number one among the alarmists, who knowingly and systematically paint a black picture of the situation. We are witnessing a global manipulation of public opinion by a body attached to the United Nations. Instead of fulfilling its role of informing and advising world opinion and policymakers, it is responsible for creating groundless fears.

There can be no doubt today that the IPCC holds the main responsibility for the great fear of climate change, which has swept across the world. In principle, people tend to trust a body run by the United Nations, which claims to group thousands of experts and has received the Nobel Prize. It is all the more regrettable that when we examine the documents it publishes – which everyone can consult freely – we have to reach the opposite conclusion.

The IPCC gives a biased view of the world, which is systematically negative and ignores the facts. Passed on by climatologists and other people who make their living in this field, the view is directly inspired by the ecologist movements, which are

73

omnipresent alongside the IPCC and even within the organization: Greenpeace, WWF, Friends of the Earth and many other bodies. They never fail to take part in its meetings, even when they are held on the other side of the world, no expense being too great for them as they are highly subsidized. This explains why a number of experts have stopped cooperating with the IPCC and denounce its excesses.

Alleged impact of temperature rises

Other 'examples of regional impacts' described by the IPCC in the Synthesis Report reinforce this impression. With endless talk of famine, drought, flooding, cyclones, diminishing harvests and heat waves, it looks like we are heading for the end of the world.

Some of the examples are so excessive that they become ridiculous. For instance, describing one of the catastrophes predicted for the end of the century, they state: 'Endemic morbidity and mortality due to diarrhoeal disease primarily associated with floods and droughts are expected to rise in East, South and South-East Asia due to projected changes in the hydrological cycle.'

Predicting the increase in the number of cases of diarrhoea at the end of the twenty-first century would be laughable if the subject were not serious. Apart from anything else, it fails to take into account the work currently done on a global scale to fight against this dreadful scourge, which will in all probability have disappeared well before the middle of this century thanks to the spread of health and hygiene measures being adopted throughout the world. Every week, 100,000 new households are being connected to efficient water sewage systems according to the World Bank reference publications.[10]

The other 'examples' of the regional impacts of climate change systematically paint the same black picture. For instance, the document states that in Africa: 'By 2020, between 75 and 250 million people are projected to be exposed to increased water stress due to climate change,' although the notion of 'water stress' is not explained.

In 2080 (why this date?), the IPCC similarly states that 'an increase of five to eight per cent of arid and semi-arid land in Africa is projected under a range of climate scenarios', concluding that agricultural production could be severely compromised in many countries, adversely affecting food security and exacerbating malnutrition. Is it really reasonable to predict changes in aridity in Africa with such precision more than 70 years in advance?

The predictions are no brighter for Asia, Australia, Europe, North America or Latin America. Of course, no mention is made at all of the research currently underway throughout the world to enhance agricultural productivity, to develop new, more resistant plants requiring less water, nor any other information which could mitigate such a dark picture of the future.

The list is endless. The IPCC's Synthesis Report speaks of nothing but catastrophes, without voicing the slightest doubt about them despite the fact that they can hardly be considered to be certainties. They seem to have forgotten the terms 'likely' and 'very likely' used in their own conclusions concerning sea levels and temperatures. When they mention practical examples of the possible impacts of climate change, they depict a totally apocalyptic situation, as if the aim were to throw people into a panic.

In a similar vein, the 2007 report stated that the Himalayan glaciers will have disappeared by 2035, or even earlier, which caused great concern among many Asian countries where the rivers flowing from these mountains play a vital role. However, in early 2010 it became clear that this was not the case and that the evidence for this claim came from a newspaper article that was devoid of any scientific value and which, for that matter, mentioned 2350 and not 2035. Although the IPCC was forced to admit its mistake, it claimed that this did not throw its conclusions into question, as if this had been a mere detail rather than a pivotal issue for much of Asia.

Shortly afterwards, the IPCC was obliged to admit that another of its forecasts, announcing the disappearance of 40 per cent of the Amazonian rainforest due to drought had no basis either. Slowly but surely, we finally need to face the facts: the IPCC does not have the scientific substance that it claims to.

Guiltmongering

The IPCC's role does not end there. After it has explained that we are heading for a disaster, it then turns to another approach, guiltmongering. The IPCC claims that the dangerous future of climate change caused by our own neglect could be mitigated quite easily.

After giving a dramatic description of the alleged impact of inaction, the report published by the 'United Nations expert group' following a meeting in Bangkok in May 2007 claimed that a reduction of 0.12 per cent per year, at the most, in the growth rate of the world economy, i.e. three per cent in 25 years, would be enough to control the trend in emissions, to reach an acceptable stabilization level for atmospheric concentrations and to limit global warming from 2050 onwards.

This time, the report was systematically optimistic, presenting things in such a way as to give the impression that there was an easy solution to the problem and thus justifying the need to make people feel guilty: 'Because the remedies are within easy reach, and many of them would even cost very little, what are you waiting for before doing something about it?'

The press ran headlines such as 'The cost of fighting against climate change is low,'[11] and the European commissioner for the Environment, Stavros Dimas, immediately declared: 'We have no more excuses for inaction.' The United Nations Secretary General Ban Ki-moon also added: 'We can combat climate change. The number of disasters has increased and we have scientific evidence that humankind is the cause.'

As for the chairman of the IPCC, Rajendra Pachauri, he unhesitatingly declared, basking in the aura of the Nobel Prize: 'We have seven years left to reverse the CO_2 emissions trend,' although this totally contradicts all the forecasts in his own country, India, which is continuing to open coal-fired power stations.

Following the meeting in Bangkok, one of the main writers of the IPCC's conclusions, Bill Hare, who is also an adviser at Greenpeace, said he was delighted by the consensus they had found. Hans Verolme, a member of WWF International also claimed: 'The scientists sent the fundamental message to the politicians that clean technologies were available to solve the problems and at a very low cost to our economies.'

Many people expressing these views do so in good faith. As they are all convinced that the growth of CO_2 concentrations leads to catastrophe, they conclude that there must be technical solutions to avoid it. However, as we saw in the first chapter, such solutions do not in fact exist. In the end, all they have to offer is a series of 'all you have to do is…'

Many experts thought that the conclusions of the meeting were overly optimistic and that an effective fight against emissions would cost far more than the announced figures, but their voices were stifled. They pointed out that the hypotheses retained in the report were systematically biased and were unrealistic. The latter were simply aimed at giving the most positive picture possible of the cost of a hypothetical transition towards a control on emissions, although in fact this would only be possible if three conditions were met: first, the adoption of a global international policy, second the immediate and massive flow of investments in new technologies yet to be discovered and third, no resistance to change on the part of consumers.

Of course, this is not how things will happen. Contrary to what the guiltmongers would like us to believe, there are no economic solutions for reducing emissions. They wrongly argue that such solutions are being opposed by hidden pressure groups driven by vested interests and indifference to the fate of the world, and which governments are afraid to combat. The truth lies elsewhere. The reason why it is impossible to prevent the growth of greenhouse gases is that four-fifths of the world's population has a dire need to use the remaining underground hydrocarbon resources in order to escape from poverty and acquire a decent standard of living, and that they will not give them up.

Before ending this passage on the IPCC, we must reveal its true nature. In the collective imagination, the IPCC groups 2,500 independent experts and it is this view that gives the organization its credibility. This is false.

It is interesting to note that the French title of the organization is the Groupe d'experts intergouvernemental sur l'évolution du climat, or literally, the Group of Intergovernmental Experts on Climate Change, whereas the English title, the Intergovernmental Panel on Climate Change, makes no mention of the term 'experts'. This is not by chance.

Despite its name in French, the IPCC is not designed to be a group of experts but of representatives of the member countries' governments, some of whom are scientists but others civil servants, such as directors of ministries, or even members of ecologist NGOs. They are all bound to be more or less dependent on the government departments that have appointed them, meaning that they are not independent.

In practice, the ministries of ecology or environment in the different member countries choose members of the IPCC. And, of course, throughout the world, these environmental ministries are, by their very nature, in the hands of ecologists. There is no chance of finding a single person among the representatives who questions the official line on anthropogenic climate change and the ruling *pensée unique*. It is like looking for an atheist in a church.

It is true that there is a network of 2,500 independent experts who publish scientific studies on the subject, but they are not the ones who define the IPCC's doctrine and draw up its conclusions as expressed in the 'Summaries for Policymakers'. The latter are entirely in the hands of a small number of representatives designated by the different member countries and the main environmental NGOs.

This develops into a vicious circle. The IPCC's conclusions and recommendations do not reflect the work of 2,500 independent experts as widely publicized, but come from a small group of a few dozen representatives of the ministries of environment or ecology in the member countries, chosen for their past positions and stances, and who are anything but neutral and independent.[12] In particular, it is this small group which draws up the only document designed for non-specialists, the Synthesis Report. Every paragraph, every line and every word are discussed in great length, and the result often contradicts the work done by the '2,500 experts', who do not have an input at this stage. What other explanation can there be for the document's absurdities and outrageous claims revealed earlier in this chapter?

The most activist government representatives – and the worst gloom-mongers – form the IPCC's elected Bureau. Its 30 members all share the same certainties and are convinced that they have the monopoly on truth. They are 'true believers', followers of the 'real faith'. As they are elected by their peers and are all convinced of anthropogenic climate change, there is not the slightest chance of a single 'sceptic' being a member of the IPCC Bureau, the system's core organ.

For instance, France is represented at the IPCC by the glaciologist Jean Jouzel, who we have already mentioned above. He is one of the intergovernmental body's twenty (!) vice-chairs and is also one of the most pessimistic. In France, his many declarations to the media and his publications constantly back up the idea that we are heading for a disaster[13]. His own work on ice in Greenland has shown, as we have seen, that the climate has sometimes undergone sudden rises of 16°C, and above all that temperature rises had always in fact *preceded* the changes in CO_2 concentrations and not the reverse. This does not prevent him from claiming that past periods of warming are 'relevant to future changes in our climate' (op. cit.), despite the fact that we have only recorded a rise of 0.7°C in the past one and a half centuries.

Appointed, in conjunction with Nicholas Stern, vice-chairman of the French Environment Summit (Grenelle de l'environnement) by the minister of ecology, Jean-Louis Borloo,

Jouzel played a key role in France in spreading an extremely negative view of the future and in encouraging France to adopt measures that are as ruinously expensive as they are ineffective. The same is true, to a greater or lesser extent, in all the UN member countries, where staunch alarmists control the official word.

It should be said that fate was sealed from the start, as the IPCC's official mission, as defined in its statutes, is as follows: 'The role of the IPCC is to assess the information relevant to understanding the scientific basis of risk of human-induced climate change, its potential impacts and options for adaptation and mitigation.' In other words, any doubts about human responsibility were excluded from the start.

The mechanics of the IPCC were closed from the outset as anyone can check by looking at the official IPCC website. Every five years or so, the IPCC produces an assessment report in a fashion that leaves nothing to chance. From the very beginning of the preparation process, IPCC officials define the outline of the report to be published five years later. In other words the conclusions *precede* the research. Subsequently, NGOs select, in consultation with governments, the experts who will produce the report. Therefore, quite unbelievably, organizations such as Greenpeace, WWF and Friends of the Earth more or less appoint the experts who are thought to be representative of the United Nations. Those in disagreement are ruled out. Under such circumstances it is evident that there can be no debate.

In its proselytising, the IPCC can also count on the unfailing support of two other United Nations bodies which play a key role in global communication.

First, the UN expresses its views on climate change via the United Nations Environment Programme (UNEP), the equivalent of a ministry of environment which, like all the others, is a citadel for those who believe in anthropogenic global warming (AGW).

It also uses the United Nations Framework Convention on Climate Change (UNFCCC), a body founded in 1992 and named after a text signed that year by 192 countries, i.e. nearly all the UN member countries. This body's explicit aim is to fight against climate change, once again leaving no room for doubt about human responsibility.

In the end, it is these three entities that guide world opinion and policy and it is therefore no surprise that their leaders get so much media coverage, even if their names are usually unknown to the general public.

The respective heads have been Rajendra Pachauri, chairman of the IPCC, Achim Steiner, executive director of UNEP and Yvo de Boer, executive secretary of UNFCCC. It is difficult to find the slightest variations between their declarations, especially as officials sometimes move from one body to the next. The three leaders all convey catastrophic messages for the future of the planet, claiming that the fault lies with humans and urgent action is required. Pachauri declares: 'I hope this report will shock people [and] governments into taking more serious action.' Achim Steiner takes a similar line: 'In the light of the report's findings, it would be irresponsible to resist or seek to delay actions on mandatory emissions cuts.' Yvo de Boer says exactly the same thing: 'The findings leave no doubt as to the dangers that mankind is facing and must be acted on without delay' (*Financial Times*, 3 February 2007).

Not able to stand the failure of Copenhagen, Yvo de Boer resigned spectacularly at the beginning of 2010. To understand this, you have to know that he had burst into tears on stage two years earlier during a preparatory meeting of the same Congress, when he realized he was not able to obtain any agreement on the reduction of global emissions of CO_2, which he himself considered possible against any evidence. This type of reaction is revealing. Yvo de Boer is, as are Achim Steiner and Rajenda Pachauri, representative of the great majority of those for whom climate change and human responsibility are a matter of faith and escape any rationality. They are convinced they have a mission on earth, which is to save our planet. The quotation of Montaigne applies to them: 'The less we understand something, the more we believe in it.'

It is interesting to note Yvo de Boer is a diplomat, as is Achim Steiner, and thus has no legitimacy to speak about a matter so complex and difficult as that of the evolution of the climate. The same goes for Rajendra Pachauri. None of them is a climatologist. They are incompetent, but they are 'believers'.

Like in a Dumas novel, the three musketeers of *la pensée unique* can also count on support from a fourth character that plays a key role.

Sir John Houghton, a mystical British man, is very well known in the UK, where he has received many awards and honours. After being in charge of a very pessimistic report on local air pollution, which later proved to be wrong, he headed the UK Meteorological Office, a body that plays a leading role in measuring global temperature trends. But he was above all one of the first chairmen and ideologists at the IPCC, where he was said to have set the guidelines for action in a brief phrase: 'Unless we announce disasters, no one will listen.' In other words, the end justifies the means and never mind about the truth.

This sums everything up. Houghton went on to declare, in line with the above precept, that: 'Global warming is now a weapon of mass destruction.' During his many years as chairman of the IPCC he worked hard to credit the idea that the world was heading for a manmade catastrophe. He succeeded, to say the least. Houghton managed to turn long-term climate change – an issue that had always existed but which only a handful of specialists were preoccupied with – into one of the world's major subjects of concern.

The success was such that his successors kept the same line. This explains, for example, why they dare to claim against all the evidence that a potential three millimetres rise per year in sea level will submerge whole countries. The recipe had worked remarkably well in the past so there was no reason to change it. As the French popular saying goes, 'the taller the tale, the more people will believe it.' Managing to make the entire world believe that a rise in sea level of three centimetres per decade, i.e. a small wavelet, could lead to the evacuation of certain Pacific Islands and of many various other parts of the world proves his skill in persuasion. The panic sparked off by the IPCC's claims is having the desired effect. For instance, the UN delegate for the Barbados Islands, whose highest point is 336 metres, recently declared: 'It is vital to strengthen the fight against climate change. We are not prepared to sign a suicide agreement that causes small island states to disappear' (*Time Magazine*, 15 December 2008).

It must, of course, be noted that not all the scientists agree with the IPCC's excesses. For example, 31,000 of them signed a petition in the United States, initiated by the University of Oregon (dubbed the 'Oregon Petition'), which pointed out that: 'There is no convincing scientific evidence that human release of carbon dioxide (…) is causing or will, in the foreseeable future, cause catastrophic heating of the Earth's atmosphere and disruption of the Earth's climate.'

Similar initiatives, supported by leading scientists, have been introduced in many other countries. The end of 2009 saw the launch in London of the Global Warming Policy Foundation, on the initiative of former chancellor of the exchequer Lord Nigel Lawson, who has been fighting steadfastly for years for the truth to be established, and against government policies that waste the money of taxpayers around the world on the illusion of 'saving the planet.'[14]

In the same vein, we must also mention an organization founded in 2007, jokingly called the Non-governmental International Panel on Climate Change (NIPCC). Sponsored by a former president of the US National Academy of Sciences and of the American Physical Society, the group is the fruit of an initiative by leading scientists who refute the official theories on global warming. This new association recently published a work, whose title could not be more explicit: '*Nature, Not Human Activity, Rules the Climate.*'

Admittedly, non-specialists cannot form a definite opinion after reading this report, but it contains enough arguments to at least cast doubts. For example, it mentions that a large number of indicators show that the period of the mediaeval climate optimum, around the year 1000, was hotter than the current period, which is confirmed by the fact that the Vikings lived for three centuries in Greenland where they cultivated cereals and even vines and raised cattle. According to the authors of the report, far more than greenhouse gases, it is variations in the sun's activity that cause climate change. In truth, this book, which includes contributions from 30 or so scholars from around 15 countries, inspires far more confidence than the IPCC's official documents with all their exaggerations and implausible arguments.

However, the IPCC and the other United Nations bodies are not the only ones to play on people's fears and promulgate guilt. They can count on a host of other people and bodies.

The 'alarmist in chief'

The first is the American climatologist James Hansen, who is currently director at the NASA Goddard Institute for Space Studies and has frequently been in the headlines for the past 20 years. On 23 June 1988, in the middle of an exceptional heat wave, he made a statement to the US Congress with catastrophic predictions for the future of the planet, claiming that it had entered a period of enhanced warming resulting from human activities, and that the outcome was impossible to predict. Exactly 20 years later, on 23 June 2008, he went still further. Returning to the hypothesis that the climate was 'nearing dangerous tipping points', he predicted that sea levels would rise by around two metres by the end of the century, i.e. an estimate that was six times higher than the IPCC's.

Consequently, Hansen insisted that global agricultural and forestry practices had to be reformed, a carbon tax introduced, plus a moratorium on new coal-fired power plants followed by a complete ban throughout the world by 2030! In his opinion, there is no doubt about who is responsible for the disaster. CEOs of fossil energy companies 'should be tried for high crimes against humanity and nature', as they are guilty of doing everything possible to hide the truth, in the same way as the tobacco companies have done in the past. Going still further, Mr Hansen, who his opponents call the 'alarmist in chief', did not hesitate to qualify the trains carrying coal to electric power plants as 'death trains' in reference to those of evil memory that crossed Europe during the Second World War.

Clearly, the fact that the extraction of oil, natural gas and coal resources has led to an unprecedented level of development for mankind in the past century, has helped hundreds of millions of human beings to escape poverty and hundreds of millions of others to survive, is of no importance in Hansen's view. And yet it is quite obvious that if we had not used the planet's fossil energy resources, mankind would now be in a dramatic situation. Even though there remain, of course, many problems to be solved, the fact is that life expectancy in developing countries rose from 27 to 65 years during the twentieth century; that the number of human beings who are properly fed rose by over two billion in the past 30 years; that the standard of living of most of the inhabitants of poor countries has doubled; and that over one and a half billion people now use the internet – all of which is official data published by the United Nations – can first of all be attributed to the availability of abundant, cheap supplies of energy. If the CEOs of the oil, gas and coal companies had refused to extract this energy from the ground, then they could have been accused of crimes against humanity.

Al Gore, a global charlatan

However, James Hansen was only the first in a very long line of alarmists. The most emblematic representative is without doubt the former US Vice-President Al Gore, who, after failing by a whisker to be elected in 2000 against George W. Bush, found a way of making headlines and money from climate change. This description may shock certain people, but not when it becomes clear that most of Gore's claims are lies. The only true part of his film, *An Inconvenient Truth*, which won him the Nobel Prize, is the introduction, which relates the increase in human-related greenhouse gas emissions in recent decades. The rest is fiction, whether it be the sudden temperature changes or the rise in sea levels of several metres, the submerging of entire towns and regions, the multiplication of cyclones and other natural disasters or the generalization of drought and famine, etc.

Some debate might still have been possible about Gore's claims had it not been for the initiative of a British father who decided to take the case to court when the British government decided to show the film in 3,500 schools. He asked the court to ban the movie on the grounds that it contained serious inaccuracies and was indeed a means of brainwashing children with propaganda.

In the High Court ruling in London in October 2007, the judge found nine major errors, which are worth quoting here as proof of blatant bad faith, as related by Serge Galam in his recent book[15].

- *Gore's Claim*: Some low-lying Pacific atolls have already been evacuated.
  Judge's Response: No evidence.

- *Gore's Claim*: The Gulf Stream, which warms up the Atlantic, would shut down.
  Judge's Response: It is very unlikely.

- *Gore's Claim*: Graphs showing the rise in CO_2 and the rise in temperature over a period of 650,000 years show an exact fit.
  Judge's Response: The two graphs do not establish what Mr Gore asserts.

- *Gore's Claim*: The disappearance of snow on Mt Kilimanjaro in East Africa was due to global warming.
  Judge's Response: It cannot be established that this phenomenon is mainly attributable to human-induced climate change.

- *Gore's Claim*: The drying up of Lake Chad is due to global warming.
  Judge's Response: Insufficient evidence to establish the exact cause.

- *Gore's Claim*: Hurricane Katrina can be blamed on global warming.
  Judge's Response: Insufficient evidence to show that.

- *Gore's Claim*: Coral reefs are bleaching because of global warming and other factors.
  Judge's Response: Separating the impacts of stresses due to climate change from other stresses is difficult.

- *Gore's Claim*: A sea level rise of up to 20 feet would be caused by the melting of ice sheets in the west of the Antarctic and Greenland 'in the near future'.
  Judge's Response: This is distinctly alarmist.

- *Gore's Claim*: Polar bears were drowning after swimming long distances to find the ice.
  Judge's Response: The only scientific evidence available indicates that only four polar bears have recently been found drowned, because of a storm.

In conclusion, the court found that 'the science is used, in the hands of a talented politician and communicator (...) to support a political programme,' and demanded that the film should not be shown in schools unless it was accompanied by a brochure designed 'to prevent promoting partisan political views'.

What was happening in France at the same time is all the more regrettable and striking as the film was sponsored by the Ministry of the Environment; the Ministry of National Education distributed thousands of copies of an 'educational' DVD based on *An Inconvenient Truth*, which, as we have seen, is actually an

CLIMATE: THE GREAT DELUSION

accumulation of lies. One of the government's chief advisers, the climate specialist Jean Jouzel who we mentioned above, was not afraid to say that: 'Everything Al Gore says is true,' although nearly everything is false.

As for Gore, he refrained from attacking the High Court judgement knowing he did not stand a chance of winning an appeal. He merely said that the nine errors they had found were very little compared with the thousands of other claims made in his film. This is another obvious untruth, as these nine errors are precisely the points that marked people's imaginations worldwide. It is hardly a small detail to claim that the oceans will suddenly rise by six metres instead of three centimetres. Gore is lying, but that does not stop him continuing to make a fortune travelling around the world, and it is perfectly understandable that Claude Allègre calls him a '*truand*' (crook).

Only people who know nothing of the way Gore lives can take his claims and recommendations seriously. The day after his standing ovation on receiving two Hollywood Oscars for his film, the American press revealed that he lived in a mansion worthy of Scarlett O'Hara and that his electricity bill amounted to 25,000 US dollars, i.e. 20 times the national average. The swimming pool alone consumes more than a standard family home. Gore could not have given a better illustration of how unrealistic his demands were if he had done it on purpose! Of course, like many proselytes of sustainable development, he travels the world year in year out by plane and helicopter, and no-one has ever met him in the underground or on a bus and probably not on a bicycle either. It is also true that he is paid 200,000 US dollars for each of his conferences, plus hotel expenses and three first class air tickets. That such a person should have received the Nobel Prize tells us a great deal about the weight of *la pensée unique* at the present time.

If there remained any doubts as to the amateurism of the former US vice-president, they were lifted by his declarations on 17 July 2008, when he announced his survival plan for the United States. As the future of human civilization was supposed to be at stake, he demanded that America stop using any oil, natural gas and coal within ten years and produce the energy required using sun, wind and other 'green' energy sources. He also added, against the most elementary evidence, that this aim was physically achievable and affordable, and even that it would create millions of jobs, although, of course, it would in itself lead to a massive recession. With his usual sense of moderation, he said that it was also a question of preventing hundreds of millions of 'climate refugees' from invading the United States.

In other words, it was a perfect illustration of ecological political correctness, which does not stop at calling on ancestral fears of invasion by hordes of foreigners, which, in other times, led people to dread the 'yellow peril'. The rich countries have always had a considerable power of attraction for people from poor countries as we can all still see today. This has nothing to do with global warming. And migration is increasingly well controlled and will be even more so in the future. The fear of seeing the developed countries invaded by 'hundreds of millions of climate refugees' is pure fiction. As we all know, it is by favouring development in poor countries, implying among other things that they have access to fossil energies, that we will help their inhabitants escape from misery.

Of course we should prepare for a world with no, or very little, carbon by the end of the twenty-first century, as most of the fossil resources will have been used up by that date. But to set the goal of reaching this objective by 2018 is a flight of the imagination and should definitively destroy people's trust in the person proposing it, who dares to state against all the evidence that the objective is attainable and who, on top of this, seems to actually believe it too. We might wonder, by the way, what kind of president of the United States Gore would have been if he had been elected.

The economists: Sir Nicholas Stern

In the face of such bleak prospects, it is easier to understand the reactions of the other key stakeholders. For instance, the leading British economist, Sir Nicholas Stern, hit the headlines in 2006 when he published a sensational report requested by the British government on the potential impacts of climate change. With a keen sense of communication and to illustrate the IPCC's catastrophic predictions – which he accepted without discussion – he declared that the cost of global warming was expected to be higher than the two World Wars fought in the twentieth century, putting forward a figure of 5,500 billion US dollars. This comparison, which is obviously not based on precise calculations of any sort, immediately brought him worldwide renown.

It is on the basis of this hypothetical cost that the Stern Report then recommended that the world community devote one to three per cent of its GDP each year to expenditure designed to fight against climate change. Spend little now to avoid having to pay more later on. In short, a form of insurance. The idea is as attractive as it is unrealistic, because the 'little' is in fact gigantic. One to three per cent of global GDP currently represents 400 to 1,200 billion US dollars and this is bound to rise in the future.

When it was published, many people paid tribute to the Stern Report – including the author of this book – as a remarkable document destined to become a reference. What can be more tempting than the idea of paying a small amount now to avoid having to pay far more at a later date?

As time passed, the judgement changed. While Stern's reasoning was no doubt logical, it did not stand up to the facts. The first error was in adopting as the gospel truth the IPCC's claims that the world was heading for the apocalypse and that there is bound to be an enormous price to pay. We have seen that this is not the case. Nothing proves that there is a link between carbon dioxide concentrations and the climate and even if it were the case, claiming that a rise in sea levels of 3 centimetres per decade would be a catastrophe and that mankind will not be able to face a possible, but far from certain, rise of two to three degrees celsius in the average global temperature, is simply not serious. Announcing that the cost of these possible phenomena would be greater than the two World Wars is even less so. It was a perfect way of attracting media attention, but it sparked off fear and gave credit to an eminently negative view of the future without any proof.

There was an even more serious error. The claim that the cost of controlling greenhouse gas emissions would not exceed a figure in the region of one to three per cent of the planet's GDP is flawed. It implies that the cost of reducing CO_2 emissions would be modest, in line with the recommendation of the International Energy Agency, which stipulates: 'None of the technologies required to reduce emissions are expected – when fully deployed and commercialised – to have an incremental cost of more than 25 dollars per tonne of avoided CO_2' in both developed and developing countries.

In certain circumstances this could be possible, but in most cases it is just wishful thinking, as shown by the small impact on consumption of the rise in oil prices from 2005 to mid-2008. The rise was close to 100 dollars per barrel, the equivalent of a tax of about 250 dollars per tonne of CO_2 emitted, since the combustion of one barrel produces 400 kilos of CO_2. The impact of this rise, which was ten times higher than the hypotheses retained by the IEA and the Stern Report, was marginal in terms of world oil consumption as global oil consumption continued to rise during this period. To reduce oil-related CO_2 emissions, it would doubtless be necessary to put the price of oil up to 500 dollars or more per barrel, thereby paralysing the global economy. This observation simply confirms that the world today cannot do without oil and that it is not possible to reduce the level of demand in a significant way. The same is true for natural gas and coal, to a similar extent.

This changes everything. The Stern Report implicitly put forward the hypothesis that an extra cost of 25 dollars per tonne of CO_2 avoided would be enough to reduce global emissions, but this is an illusion. The cost would have to be incomparably higher to be effective. It is not 25 dollars that would have to be spent on average to avoid emitting a tonne of CO_2, but several hundreds of dollars, meaning that the operation would not cost one to three per cent of annual global GDP, but perhaps 10 or 20 per cent. The cost would then far outweigh the possible benefits. It must be added that Nicholas Stern adopted an unrealistic low discount rate to back his reasoning; otherwise, he would also have come to very different conclusion.

Stern is an economist. He is not a climatologist, nor a specialist in energy. The theory he put forward does not stand up to a study of the facts, and his report misled the entire planet. He was unable to resist the strong convictions of the numerous alarmists in his country, such as John Houghton, and indeed he became one of them, assuming he was not one before. The author failed to confront the economic theory with a few common sense questions. Were the figures on the cost of global warming put forward by the IPCC founded? Was it realistic to think that we could massively reduce the world's oil, natural gas and coal consumption and leave significant quantities of these products in the ground unused? What exorbitant levels of tax on energy products would have to be inflicted to obtain such results?

If Sir Nicholas Stern had asked himself these questions, he would doubtless have produced a very different report.

The International Energy Agency

The International Energy Agency (IEA) is an autonomous body operating within the framework of the Organization for Economic Cooperation and Development (OECD), which groups together all the developed countries. Headquartered in Paris, it brings together a number of experts from the developed countries and is seen as the most competent organization for energy-related matters. It publishes an annual *World Energy Outlook*, reference document reviewing the current state of the planet in this domain.

In 2005, the IEA was asked by the heads of the world's eight leading countries to contribute to their action plan on 'climate change, clean energy and sustainable development' for their meeting at Gleneagles in Scotland. The IEA was asked by the G8 to advise on alternative energy scenarios and strategies aimed at a 'clean, clever and competitive energy future'. It was in this context that the IEA also made a certain number of recommendations for the G8 summit in Hokkaido, in June 2008 (IEA, Work for the G8 – 2008 messages).

However, after reading these recommendations, we are left with a feeling of uneasiness. The IEA had previously studied two scenarios. The first, so-called reference scenario, realistically predicted that CO_2 emissions from fossil fuel use would continue to grow in the coming decades, with emissions increasing from their current rate of 30 billion tonnes to 42 billion tonnes in 2030 and to 62 in 2050, taking into account current and planned projects in developing countries.

The second, so-called alternative scenario, tried to show global emissions could be controlled – at the price of considerable, and in fact unrealistic, efforts – limiting them to 32 billion tonnes in 2015 and to 34 in 2030.

However, the G8 countries charged the IEA with an impossible mission: to elaborate, in line with the IPCC's requests, a third scenario whereby global CO_2 emissions are at least halved by 2050 compared with current levels, i.e. reducing their energy-related content to 15 billion tonnes in the middle of the century, or four times less than the reference scenario.

The IEA was then confronted with the choice of either criticizing the IPCC and telling the truth, i.e. by saying that this scenario is completely unrealistic, or of setting out the conditions to be met in order to obtain the results and pretending to believe in them. It chose the second option, although it drew up the list of conditions to meet the G8 goal in such a way as to make it obvious that it cannot be reached.

We can judge for ourselves. According to the IEA, the conditions that must be met so that emissions are halved by 2050 instead of doubling are as follows:

- To take strong, immediate action in 2008 (this did not happen).
- To stabilize emissions by 2015 at the latest, although no mention was made of how to go about this.

- As of 2012, to generalize carbon capture and storage techniques (CCS) for new thermal electric power stations (in another IEA document (*World Energy Outlook 2006*) where it states that these techniques will not be ready before 2020 at the earliest, and even this is not at all certain and would be extremely costly).
- To close down at the same date 15 per cent of existing thermal power stations that do not have CCS (i.e. all of them).
- To double the number of nuclear plants by 2030 compared with the reference scenario (833 GW compared with 415), although the IEA claims elsewhere that political and economical obstacles make this objective unrealistic.
- To double electricity production using renewable energies despite the high additional costs involved.
- To invest 7,400 billion dollars in new installations in the energy sector, which would lead to considerable increases in the price of electricity for consumers.
- To drastically reduce world consumption of petrol, natural gas and coal despite pressure from demand.

These points make it clear that the target set under pressure from the IPCC and unwisely adopted in July 2008 by the G8 heads of state and government is completely utopian. This is already true for the developed countries, but how can we force the rest of the world, which consumes five times less energy per inhabitant on average, to cut down still further? It should be added that it would involve stabilizing atmospheric concentrations of CO_2-equivalent at 450 ppm, or 360 ppm of CO_2, which is less than today!

The IEA concluded by underlining the very substantial cost of carrying out this scenario. The cost per tonne of CO_2 avoided would be relatively affordable up to 15 billion tonnes less than in the reference scenario. From 15 to 30 billion tonnes less, they would amount to 50 dollars per tonne of avoided CO_2. Beyond 30 billion, the cost per tonne avoided would rise dramatically to 100, 200, 500 dollars or even more, resulting in expenditure of thousands of billions of dollars every year. This is even more unrealistic when we consider that a large part of the cut-downs would need to take place in developing countries and thus would almost certainly have to be paid for by them.

Finally, we should remember that, to be effective, the final aim of this expenditure would be to leave the greater part of the world's remaining reserves of oil, natural gas and coal unexploited.

However, doubtless frightened by the IPCC's apocalyptic forecasts should atmospheric CO_2 concentrations continue to grow, the IEA officials, who are not climate specialists, did not think they were entitled to question them. This was how the vicious circle was created.

In other words, the IEA failed in its mission. It did not tell the G8 heads of state and government the truth, that is that their demands, based on those made by the IPCC, were impossible to meet. It also failed to clarify the situation. By using esoteric units such as ppm and Gt in its reports, it made them incomprehensible to ordinary people and to the politicians.

In 2009, the IEA moved even further towards renouncing its independence. The body actually entrusted the preparation of a chapter of its annual report to Yvo de Boer, the then executive secretary of the UNFCCC and one of the most ardent propagandists of the official line. So the IEA dares to declare that emerging countries should not increase their CO_2 emissions in the future, even though levels currently stand at no more than 1.4 tonnes per capita per year, China excluded.

When one bears in mind that average emissions from rich countries amount to 15 tonnes per capita, one is lost for words. What justification can there possibly be for imposing targets on the inhabitants of poor countries that would condemn them to remain in the most abject poverty? It is particularly fortunate that it is these countries, rather than the IEA, that decide their own policies.

The G8

We can hardly blame the G8 heads of state and government and instead should try putting ourselves in their shoes. Firstly, the leaders of the eight countries received the apocalyptic predictions drawn up by the IPCC, which had just been awarded the Nobel Prize; secondly, they were given reports from the IEA that at a first glance seemed to suggest that it was possible to control global emissions, although this is not the case. The Stern Report said the same thing. In all honesty therefore, they believed that they were doing the right thing in setting ambitious global targets regarding CO_2 emissions, without realizing that these were absurd as they implied that extraction of oil, natural gas and coal should be drastically reduced. We should therefore not be surprised by the 'decisions' taken by the G8 in Hokkaido which, flying in the face of common sense, plan to halve global greenhouse gas emissions by 2050 compared with the current level (and without specifying the reference year).

The G5 countries (China, India, Brazil, Mexico and South Africa) understood this and refused to commit themselves to any such target, which they believed would unacceptably hamper their development. In fact, in order to continue developing they need to increase their energy consumption significantly, so their greenhouse gas emissions will continue to grow for many decades to come instead of declining. After reading the IEA's reports and realizing that the global target they set for 2050 was unrealistic, it is easy to understand why the G5 called the developed countries' bluff and asked them to begin by committing themselves to targets for 2020 instead of such distant dates. Their reasoning was that to reach the ambitious targets set for 2050, a sudden change would have to be made straight away and a first stage completed to obtain an immediate reduction in emissions from rich countries: a goal the undeveloped countries knew to be impossible.

The developing countries also put forward moral arguments. Although they now represent nearly half of global annual emissions, in the past it was the developed countries that made the most contribution to the growth in global CO_2 stocks in the atmosphere and even now they continue to emit on average five times more CO_2 per inhabitant than the poor countries. It would therefore be up to them to take action instead of simply announcing far-off targets that do not commit them to anything. With the exception of Europe, the rich countries did indeed refuse to commit to any targets whatsoever for 2020, particularly the target demanded by the G5 of reducing their emissions by at least 25 per cent compared to 1990 levels. They were in fact quite right to refuse, as a reduction of this scale is inconceivable, except if there were to be a global cataclysm, which of course nobody could want.

The problem was therefore postponed until the end of 2009, when leaders were supposed to come up with a follow-up to the Kyoto Protocol at the Summit Meeting in Copenhagen. It was doomed to fail. As we have seen, the truth is that there is no solution to reducing global emissions. This is simply not possible, given that works are continuing all over the world to increase the production of hydrocarbons, and that the fossil energies which certain people fail to consume will simply be consumed by others.

The pressure groups

Lastly, we must mention one of the major causes of the absurd ideas currently held about climate change and the assumption that humans are responsible for it. This is the influence of pressure groups, which can be divided into two main groups: first, the ecological movement and second, the enterprises whose business depends either wholly or partly on public funding supposed to help 'save the planet'.

In France, the Grenelle Environment Summit (Grenelle de l'environnement) provided a striking example of the causes of disinformation, as it clearly showed that pressure groups had a stranglehold on the government and controlled its decision-making processes. The groundwork was done during the summer of 2007 by a working party of 50 people chaired by Nicholas Stern and Jean Jouzel.

It is interesting to study the make-up of this working group as it explains everything. There were several different components, including the ecologist NGOs movements and various associations such as the trade unions and other public bodies with specific interests to defend and finally, the Medef (French employers' federation), which was represented by its leading federations (construction and public works, railway industry, wind power industry, car manufacturers, etc.).

Anyone who may have been opposed to finding a consensus was systematically excluded, such as the anti-wind farm associations, the road users' representatives and many others, although they had asked to take part in the debate. It is therefore hardly surprising that the Environment Summit came up with unanimous recommendations that allowed Nicolas Hulot, a famous media commentator in his country, to celebrate the 'unexpected' alliance of the ecologist movements and the business world.

There was in fact nothing unexpected about it at all. It was simply a case of the different parties' interests joining together, to the detriment of consumers, tax-payers and the general interest. The Medef's leading federations were obviously thrilled at the prospect of building thousands of kilometres of railways and undergrounds, insulating buildings, covering France with wind farms and solar photovoltaic panels, etc., at the cost of hundreds of billions of euros, as it is in their interests to continue to develop their businesses. The only problem is that it will be impossible to pay the staggering bill resulting from the decisions taken at this French Summit. The bill amounts to around 20 billion euros per year for the taxpayer, to which should be added an equivalent sum to be paid directly by the consumers. Their purchasing power will thus be reduced by a total of 40 billion euros per year. In most cases, there will be no justification for such spending as it will have no impact on global greenhouse gas emissions because of the global nature of the phenomenon, as we saw earlier.

Does this mean that we cannot answer the fears preying on people's minds?

3   The great fears

The beginning of this century has been a time of unprecedented progress for humanity. Up until the financial events of 2008, the growth of the global economy had settled at around 5 per cent per year, a hitherto unknown level, and hundreds of millions of people have emerged from poverty. But our era is also one of great fears. Three of these are now in the forefront of people's minds:

- Faced with the expected exhaustion of oil resources, can world expansion last, rich countries maintain their lifestyles and their inhabitants continue to travel?
- Can poor countries, whose populations represent four-fifths of humanity, achieve a better standard of living one day, which presupposes, among other things, access to electricity?
- Lastly, what will be the consequences of changes in the Earth's atmosphere we have seen to be inevitable?

I – The post-oil world

Should the inevitability of oil supplies running out lead to pessimism? Will we have to leave our cars in the garage, our planes in the airports and our ships in port as many people fear? According to the polling organisation IPSOS, almost two-thirds of French people (65 per cent against 35 per cent) think we will soon have no petrol for our cars[16].

They would certainly be correct if nothing were to change, since oil products play an essential role in the world economy and in particular for transportation that depends upon them. Fortunately, however, the prospects opened up by technical progress are so vast that changes will take place, as can be demonstrated by an inventory of the different means of transport, beginning with the car which in the space of half a century has become one of the major elements in the prosperity of developed economies through the considerable savings in travel time it allows and the opening-up of markets it engenders.

The car

If there is one fear everyone, or almost everyone, has felt at some time, it is that of finding themselves deprived of their car. Running out of petrol is one of the worries that most preoccupies people at the beginning of this century. What will happen when the Chinese, the Indians, and more generally, the inhabitants of the rest of the planet each want their own car, as is the case in developed countries where the number of vehicles now far exceeds the number of households?

Those who ask this question often imagine we will never experience such a situation because the equivalent of three to four planets would be needed to satisfy everyone's demands. Yet things will not happen this way. No doubt, the Chinese, Indians and others will have their cars one day just as we do. But those who predict we are heading for disaster are forgetting one thing: the existence and speed of progress which will allow us to own cars that consume less and less and will later use fuels or energy sources that will be substitutes for petrol when oil has run out.

From 10 to 5 litres to 100 kilometres

The first step is the easiest to achieve and has already begun. At world level, the average consumption of private cars is now in the region of ten litres to 100 kilometres. However, those sold today in France and other European countries already run on half that. It is the Americans, and to a lesser degree the Germans, who are mainly responsible for such a high world average level of consumption.

United States: The SUV vs. common sense

Until very recently, North Americans usually had their hearts set on real monster vehicles, driven by 8, 10 or 12-cylinder engines that guzzle between 15 and 20, sometimes even more, litres for every 100 kilometres. These are four-wheel drive SUVs (Sports Utility Vehicles), occasionally needed in rural areas, but more usually used to transport one or two people in suburbs with excellent road networks. In other words, they render the same service as any car sold in Europe, including small cylinder ones that now only require around four to five litres to 100 kilometres, or even less.

But things have been changing across the Atlantic since the traumatic events of 2008. Buying such uselessly greedy vehicles was the consequence of a price at the pump that had long remained very low as a result of the absence of specific taxes such as those in Europe. Petrol was almost free, so the rate of consumption was not a criteria of choice, which explains why cars were bought which would have been unsaleable in the Old Continent.

Sudden jolts in prices at the pump in the United States, the consequence of violent fluctuations in the international price of oil, have upset the established order. At the beginning of 2008, SUVs, whether new or second-hand, became unsaleable in the States, while sales of fuel-efficient vehicles have suddenly rocketed in a market that became generally very sluggish. Although after its peak in mid-2008 the oil price per barrel plummeted as suddenly as it had risen and future fluctuations remain unpredictable, people became more cautious and many believe prices will never be the same again. Has not the Smart car, a two-seater unique in its kind, seen sales take off for a while in a country for which it was not originally intended?

In July 2008, the president of the Ford Motor Company, hitherto champion of the large-engined cars that made up most of its sales and profits, announced that the company was to cease production of this type of vehicle in three of its factories and convert them to manufacture European-type cars designed by its European subsidiaries. The granting of massive federal help to the three American manufacturers came with the same requirement.

German uncertainty

Though to a far lesser degree, Germany is the other guilty party responsible for the high average consumption of the world automobile fleet. The German economy is based largely on the prosperity of its automobile industry which is by far the largest in Europe. So German automobile manufacturers have hitherto held successive governments, regardless of political orientation, in their power, even those that included Green Party ministers.

This is how Germany has come to be the only country in the world to maintain sections of motorway with no speed limit despite a high accident rate which has not been made public, in order to allow its manufacturers to turn this into an advertising argument. As a result, they have long specialized in producing increasingly more powerful and faster vehicles that have been exported all over the world, despite far stricter international speed limits. Was not the average maximum speed of a BMW, Mercedes, or an Audi still 235 kilometres an hour in 2008?

Yet each time a French government has tried to convince its German counterpart to fall into line with the world practice of generalized speed limitations, for reasons of safety and energy saving, or has suggested forcing manufacturers to produce cars with predetermined maximum speed as is the case for lorries in Europe, the Germans have refused. The unambiguous message that has come back has always been the same: if France puts this subject on the agenda, it will be a *casus-belli* between the two countries.

The opposition of Germany to speed limited cars explains why it has been impossible in the past to align European production of automobiles to those cars that consume the least. Until recently, an endless race has been run for ever increasing power, resulting in both unnecessary accidents and a high fuel consumption. What is the point of producing vehicles with an optimum speed of 200 to 250 kilometres an hour, when speeds are limited elsewhere in the world to around 120 to 130 kilometres an hour or less, and when petrol is becoming scarce?

But things are changing. In its crusade against CO_2 emissions, the European Union is requiring the continent's manufacturers to comply with draconian norms, which the firms are fighting to delay, but which will be imposed on everyone before too long. The average emissions of vehicles put on the market by each manufacturer will soon not exceed 120 or 130 grams of CO_2 per kilometre travelled, which corresponds to 4 or 5 litres to 100 kilometres.

This is a very restrictive objective when we know average emissions of cars sold in Europe stood at 164 grams of CO_2 per kilometre in 2006. This is particularly true of German manufacturers, specialists in large engines, whose vehicles emit an average of 180 grams compared to less than 150 grams for two French groups and an Italian one. German manufacturers successfully campaigned for these objectives to take account of the weight of the vehicles. But let there be no mistake. Though late in the day, Germany has now done a U-turn as almost all its manufacturers are now producing vehicles that are more efficient.

By wishing to fight against the planet's CO_2 emissions – an illusory objective – Europe will nonetheless have achieved something useful: the reduction of its petrol consumption and the consequent saving of a scarce resource.

In addition some countries have introduced a *bonus-malus* system that allocates subsidies to low consumption vehicles, and taxes high consumption ones. This decision by the government proved effective in France in directing consumers towards economic models and away from the bigger consumers of petrol. In the course of the first four months of 2008, the percentage of economic vehicles (less than 130 grams of CO_2 emission per kilometre) sales increased by a third in France, whilst that of the least economic (more than 160 grams) fell by 40 per cent compared to 2007. France is therefore now one of the planet's 'good pupils' in this respect, mainly as a result of this measure, which may be the only justified element in the national ecological policy.

One simple conclusion emerges from a comparison between the United States, Germany and France. For the moment at least, a country, France in this case, shows that it is possible to satisfy, with no restrictions whatever, individual mobility needs with cars that only require five litres or less to 100 kilometres, and when the average consumption of the world fleet is currently almost double that figure, the global short-term potential gains are as considerable as they are easy to evaluate. With the same quantity of fuel, it will soon be possible to have twice as many cars on the world's roads as today, that is to say two billion instead of about one, or to have the same number of cars and to consume half the quantity of fuel.

No technical revolution is needed since in recent years automobile manufacturers have made amazing progress, of which the general public is not always aware, and the fruits of which are only now beginning to appear on the market. Average-sized European family cars now frequently have a range of more than a thousand kilometres, a performance that was unimaginable only a short time ago.

All now depends on the policies adopted by various countries around the world. With a combination of very high taxes on motor fuel, and the *bonus-malus* system, France, and other countries as well, are showing what it is possible to achieve.

Not everyone is following, however. Until recently, China, India, Indonesia and many other Asian nations not only did not tax motor fuels but supported them with massive subsidies because they rightly judged that cars and lorries had an essential role to play in their economic development. However, it is also true that, on the whole, low living standards have spontaneously limited the purchase of cars that unnecessarily consume energy.

At global level, a lot will depend on the United States, which alone uses one quarter of the planet's motor fuels yet represents less than five per cent of its population. If it does not levy a significant tax on oil products as in Europe, it is to be hoped that President Obama will, as has already been the case in the past, require manufacturers to market vehicles with lower fuel consumption than at present. It is, moreover, significant that the new president's first decision, only the day after his election, was to announce that he would help the American automobile industry with a large-scale conversion in this direction.

When we know that automobile fleets are renewed every six to fifteen years, depending on the country, we can see how large the margins are for global reductions in the average consumption of automobiles in the short term, without any radical technical changes. But it should be possible to do much better before very long.

From five to three litres to the 100 kilometres
Given recent progress in consumption, one might have thought we had reached the limit of our technological capabilities. But nothing could be further from the truth as an official French report, named after its coordinator, Jean Syrota, shows. Produced under the aegis of the Centre d'analyse stratégique and the Conseil général des Mines, this comprehensive document is entitled *Perspectives concernant le véhicule grand public d'ici 2030 (Prospects for mass-market vehicles between now and 2030)*. Written in collaboration with the automobile industry, the report shows that current thermal petrol and diesel vehicles can still reduce their consumption dramatically if they are optimized.

Major progress is in fact possible in many areas, starting with engines. For one thing, current vehicles are more powerful than necessary, and reducing their maximum speeds to around 140 kilometres/hour would reduce at once their overall consumption by 15 per cent. Then, many other improvements could be made to the vehicle itself: lighter, better aerodynamics, tyres, etc.

The French Report concludes that:

> Propulsion by thermal engines fuelled by liquid carburant under normal conditions will remain the norm between now and 2030, particularly since their substitutes do not offer, performance for performance, significant advantages

in terms of cost, energy balance and overall emissions of greenhouse gases, neither currently nor in the long-term. Current vehicles possess a wide margin for improvement, *so a reduction by half in consumption is perfectly possible.* In particular, small thermal vehicles, with limited power (that is to say with on-road performances comparable to those of most of the coming electric vehicles) and therefore with greatly reduced consumption, could obtain a significant market share.

Whilst vehicles now coming onto the market in France consume an average of five litres to 100 kilometres, technical developments will enable this to fall to around three litres within the next decade or two without any major upheavals. It is simply a matter of harnessing the research carried out by the tens of thousands of engineers and technicians in the automobile industry and by adapting the fleet to the reality of present-day needs, as drivers have already begun to do by buying more modest and less powerful cars.

But progress will not stop there.

From three to one litres to 100 kilometres
The sudden rise in the price of the barrel between 2006 and mid-2008, together with the certitude of seeing the oilfields dry up in the future has dramatically accelerated research of all kinds, aimed at reducing the unit consumption of cars, or equipping them with engines that do not require oil products. But this has not been without problems.

Hydrogen fuel cells: a solution for the future that is destined to remain so. One might suppose that industrial sectors, which abound in supposedly rational minds, would be immune to fads and fashions, but this is not at all the case. Three or four decades ago, opinion was quasi-unanimous in asserting that if the day came when the internal combustion engine had to be replaced by something 'clean', it would be by a fuel cell. A fuel cell causes hydrogen and oxygen to react in a controlled manner to produce electricity and water vapour. This would then represent the advent of a new economy where hydrogen would replace fossil hydrocarbons.

But even supposing it were one day possible to store and transport a gas as potentially dangerous as hydrogen, enthusiasm for the latter left one basic question unanswered. Where would it come from? There were really only two solutions. It would either have had to be extracted from fossil hydrocarbons, which would have only transferred the problem, or be obtained from water by electrolysis, an operation that uses a large amount of electricity, and as we have seen, this now comes mainly from power stations run on coal and which are therefore heavy emitters of CO_2!

But fashions come and go. Thirty years later, and even though certain manufacturers are still producing very expensive prototypes, no-one claims that this will be a realistic solution before many decades have passed at the earliest, since the physical and economic obstacles are considerable.

The electric car: dream or reality?
At about the same time, electric engines were already the subject of similar enthusiasm. French manufacturers even began to produce electric cars on the same assembly lines as petrol or diesel vehicles. But the market did not take these up for reasons that are easy to understand and have nothing to do with the engines themselves, which are economical, reliable and compact. Instead it was because it is extremely difficult and costly to stock electricity. *Weight for weight, a kilo of battery usually contains 50 times less energy than a kilo of petrol.*

After being abandoned for 20 years, the recent rise in oil prices has placed the electric car back on the agenda and given the green light to a frenetic burst of new research to develop more efficient batteries with greater energy capacity that can be recharged in just a few minutes instead of hours. The development of compact, rapidly rechargeable and economic batteries is one of the major challenges humanity could be confronted with in the twenty-first century, and has been compared to the challenge met by the Americans of landing on the Moon. But there is nothing to indicate the challenge will be met.

Contrary to frequent affirmations by the ecologists, it is not a lack of goodwill that has prevented scientists and industrial firms, who have been trying since the invention of the electric battery by Volta in 1800 to develop efficient batteries, from meeting with success so far, but rather the laws of physics.

Manufacturers all over the world are seeking the best solution. The three 'big' American firms (General Motors, Ford and Chrysler) have recently begged the federal government to grant them massive research loans to help them come up with efficient batteries, asserting that it is vital for the American economy to reduce its dependence on oil.

For the moment, attention is focused on lithium-ion batteries because they are able to store two to three times more electricity than other kinds. Moreover, lithium-ion batteries are not the only possibility. A General Motors laboratory is currently exploring, aided by a computer, the structure of the 30,000 mineral products discovered by chemists in the course of the last two centuries in order to detect and test those that could present characteristics favourable to the creation of ultra-efficient batteries that can be recharged in just a few minutes or even seconds, and can at the same time store large quantities of energy in a small space. This programme has been dubbed the 'materials genome'.

In late 2008, a Chinese company, BYD announced it had developed a 'lithium-phosphate' battery providing the ability to run a light car for 400 kilometres and that could be recharged in three hours. That would be a revolution, but most specialists are sceptical.

No one can guarantee the success of the research undertaken, and let us not be confused: the batteries of the future will almost certainly never have the same capacity to concentrate energy as liquid fuels.

However, those which are near the market, such as the lithium-ion batteries, could be used for two different types of vehicles, the purely electric ones, and the others, the 'rechargeable hybrids', that might be qualified as 'semi-electric'.

'All-electric' vehicles

Cars that are solely electric constitute the first avenue of research and at least one of the world's manufacturers has announced its intention to launch such vehicles on a large scale in one or two years hence with two distinct markets in view.

Small countries

In small countries, long-distance routes are naturally few, or even physically impossible. This is the case, for example, in Israel, Ireland, other island countries in general, and to a lesser degree Portugal and the like.

In these countries, the reduced range is most often sufficient, and it is easier than elsewhere to envisage the implantation of a dense network of charging terminals or battery exchanges.

This is how an agreement came to be signed between the French automaker Renault and an Israeli firm for the future marketing of 'all-electric' vehicles in Israel and for equipping the country with a network of charging terminals where an empty battery could be exchanged for a charged one.

But in the face of competition from traditional petrol and diesel cars, will such a solution be economically viable as long as the cost of petrol has not yet attained a permanently prohibitive level? This is highly doubtful to say the least when we know that the batteries to be 'exchanged' will probably weigh around 200 kilos.

Vehicles for use in town

All over the world, there are vehicles that are not intended to travel long distances because they are exclusively for use within an urban zone where journeys are counted in tens and not hundreds of kilometres, and where batteries can be charged every night. These include not only delivery vehicles, but also private cars whose owners possess at least one other traditional vehicle suitable for travelling long distances.

Admittedly, this is a niche market, but it is one that interests many manufacturers who see in it an opportunity to pioneer new techniques.

Nonetheless, we must not deceive ourselves. The maximum we can hope for today is a battery life of around a 100 kilometres with, in addition, modest performances, and then only if the terrain presents no particular difficulties. And as things stand at present, the cost of the batteries needed to achieve such modest performances is considerable, to the point of doubling the price of the vehicle. The people who believe this may be a significant market are therefore deluding themselves. However, it must be emphasized that in this field, as in others, rationality often takes a backseat to the impulsive enthusiasms of leaders.

We can therefore understand the Syrota Report when it declares that:

> The electric vehicle, which has the advantage of not emitting polluting gases directly (but which can indirectly emit far more than a thermal vehicle if the electricity is produced in a traditional power station) suffers from too many handicaps to be a large-scale substitute for the petrol or diesel vehicle.

But this common-sense argument doubtless did not please everybody, in particular ecologists who want at all costs to make people believe their present vehicles are 'dirty' and that the electric vehicle is clean and is 'the future'. Otherwise, how else can we explain that this French report has still not been officially cleared for publication, even though it is quoted in the press and it can be consulted on the Internet by anybody. Those who oppose the publication of an official technical report because they do not like its content have a strange idea of transparency, to put it mildly.

Rechargeable or 'semi-electric' hybrids
This is why the use of electricity to propel automobiles will mainly take a different form in the coming decades.

We have all heard of hybrid cars, but a great deal of confusion surrounds them. Hybrids are vehicles with at least two engines, a traditional petrol engine, and an electric one. However, contrary to a widely-held belief, current hybrids cannot travel on their electric engine alone. The batteries they are equipped with give them, in 'all electric' mode, only *one or two kilometres* range at the most. The truth is their battery has to be constantly recharged by the petrol engine, which is running all the time. The advantage of this system is therefore not, as is commonly imagined, that the car can run on its electric engine alone, but simply that the use of the petrol engine is optimized. In case of sudden acceleration it continues to turn over at the same rhythm with the electric engine temporarily taking over to propel the vehicle. Average consumption is therefore reduced, in particular in town, but in the end, savings are modest.

The 'rechargeable hybrids' that will shortly appear on the market are a very different proposition. Provided that hybrid cars have one or even several electric engines alongside the traditional internal combustion engine, it is perfectly possible they could be equipped with more efficient batteries than at present and be able to charge these simply from electric plugs, at night for example, with the aim of giving them the ability to run for several dozen kilometres solely on electricity. And most cars do not travel more than this on a daily basis. Daily journeys could therefore be made without using liquid fuel, which would be reserved for long journeys, reducing average annual consumption to one or two litres to 100 kilometres.

It is not surprising, therefore, that the Syrota Report affirms that 'the rechargeable hybrid vehicle which combines the advantages of thermal and electric technology without any of the major disadvantages has every chance of being the vehicle of the future'.

The first prototypes of these rechargeable hybrid vehicles (known as 'plug-ins') are soon coming to the market in the United States and in Europe. General Motors has announced it will put a vehicle of this kind on the market in 2011. Christened 'Volt', it will be able to travel 100 miles to the gallon under average conditions, which corresponds to two litres to 100 kilometres, an astonishing performance when we consider that the average American vehicle now needs ten litres. Toyota, a pioneer of first-generation hybrid vehicles, doubtless also intends to be present on the 'plug-in' market in the coming years, like Peugeot and most of the world's manufacturers.

There is no doubt therefore that these 'semi-electric' vehicles will soon be on the market. On the other hand, since the fall in the price of hydrocarbons in the second half of 2008, it is difficult to know when they will be in widespread use, and their market share in years to come will depend directly on fluctuations in the price of the barrel, since the considerable cost surcharge they imply will only be justified if oil remains permanently dear, which is sure not to happen for very many years yet.

And, when the far-off day of the 'plug-in' vehicles arrives as a strong presence on the market, production of electricity current will have to keep pace, which will mean stepping up electric power plant construction programmes.

Be that as it may, one conclusion may be drawn which changes the vision we might have of the future of the automobile. In the long term, vehicles could technically consume scarcely more than one litre to 100 kilometres compared to almost ten on average at the present time, which means that all the world's inhabitants could possess a car, contrary to the forecasts of those who affirm that it would take several planets to meet demand, for they forget the existence of technical progress.

Lorries, ships and planes

Whilst we can expect real technical revolutions in the motorisation of private automobiles, nothing of the sort seems to be on the horizon for lorries, boats and planes. Admittedly, progress will continue and with it reductions in unit consumption. Rates of approximately 20 per cent improvement in ten or 20 years have generally been predicted for each of these modes of transport, but the progress of trade, in particular international trade, is far faster on average. Is it not sometimes five per cent a year in the case of air transport?

Will we see the development of propeller planes for short journeys, as they are more fuel-efficient than jets? Perhaps will we see heavier lorries that are less energy consuming on European roads, as is the case in the United States? Perhaps we will see nuclear-propelled merchant ships? Yet all this will scarcely change the scale of needs. Whatever the efforts of the engineers, we must be prepared for an increased and incompressible demand for fuel to transport goods and people by air, sea and road, which inevitably poses the question of how we will meet such needs when the oil has run out, even if it is not for tomorrow.

Synthetic fuels

Apart from oil obtained from traditional sources or from sources such as oil shale or sand, various raw materials can be used to make synthetic liquid fuels suitable for transport needs.

Coal and natural gas for example, can be used. The method currently favoured for obtaining liquid fuels is the Fischer-Tropsch synthesis, named after the German chemists who were the first to obtain such fuel using coal. The method was widely used by the Germans during the Second World War to provide fuel for the tanks and planes of Hitler's armies. More recently, South Africa, the object of a world embargo at the time of apartheid, relied on the same technique to produce the liquid fuels it needed, firstly from coal, then using natural gas, with the result that it is now the world's leading producer of synthetic fuel, with 10 million tonnes a year[17].

In the long run, if the price of petrol remains permanently high, the production of coal-based synthetic fuels (CTL, Coal to Liquids procedure) will be developed, and China is already heavily committed to this. The country's biggest mining company, Shenhua, plans to manufacture 30 million tonnes of coal-based synthetic fuels by 2020. Other projects are under examination and if successful, China is set to produce 70 million tonnes of liquid fuels each year from 210 million tonnes of coal.

Other players have developed different technologies and the beginning of the 2000s marked an important turning-point with the announcement of a major project in Qatar to manufacture synthetic petrol from the North Dome gas deposit, which is the biggest in the world (GTL: Gas to Liquids procedure). The project was launched with the participation of Qatar Petroleum, the South African company Sasol and the American company Chevron. A first unit of 34,000 barrels a day (that is around 1.7 million tonnes a year) saw the light in 2006. Other projects, currently under examination, represent a capacity in the order of 40 million tonnes a year for Qatar alone.

All things considered, however, the possibilities for manufacturing synthetic fuels from hydrocarbons should not be overestimated. They are real, admittedly, but in the face of the progressive extinction of reserves of natural gas and even later of coal, they alone will not be able to cope with meeting our need for transport when oil supplies have been exhausted. Furthermore, the sudden drop in the price of petrol in 2008 will doubtless put the brakes for the time being on a number of projects.

Biofuels

Like synthetic fuels, biofuels, that is to say fuel extracted from vegetable matter, have the considerable advantage of resembling petrol and needing no significant transformation of the supply chain or of the engines that use them. But if they are to play a fundamental role, they must be produced in large quantities, which leads us to make a distinction between the current situation on the one hand, and future prospects on the other.

The current situation: the cart before the horse
Under pressure from the ecologists, the world has arrived at an impasse. Current techniques for producing biofuels for road transport in temperate countries display several major disadvantages. Either it is a question of the alcohol known as ethanol obtained from the sugar contained in beetroot and cereals, or of an ester known as VOME (vegetable oil methyl ester) obtained mainly from rape and sunflower oil, which can be incorporated into diesel fuel up to a level of 5 per cent without adapting the engine.

Global production of ethanol, a product similar to petrol but containing less energy per litre reached 45 million tonnes in 2007, mainly in Brazil and the United States, and that of VOME, 4 million tonnes, mainly in Europe. The quantities involved are therefore marginal compared to the 4 billion tonnes of petrol produced in the world.

In European countries, there are major drawbacks to the production of biofuels as cultivated today. Firstly, their yield is very low and at best does not exceed one tonne of 'petrol equivalent' per hectare compared to four for sugar cane in Brazil, for example. The surface areas needed to obtain significant quantities of fuel are therefore prohibitive. *To reach a level of production representing 10 per cent of the consumption of fuel for road vehicles alone, it would be necessary, in Europe as in the United States, to exploit a surface area equivalent to 30 to 40 per cent of agricultural land.*

The conflict with food needs is therefore obvious as recent figures from the Food and Argiculture Organisation (FAO) show: the number of human beings suffering from malnutrition rose sharply from 850 to 925 million between 2006 and 2007 and increased even more in 2008 and 2009, the direct consequence of the rise in the price of agricultural products, engendered in part by the allocation of farming land to the production of biofuels. The situation has been especially noticeable in the United States where the American government decided to heavily subsidize maize-based bioethanol, an operation all the more open to criticism since its yield is disastrous. Certain studies have even shown that the result was negative from the energy point of view since the quantity of petrol needed to plough, sow, manufacture the fertilizer, harvest and process the maize was of the same energy value as the ethanol produced!

The very low yield of first generation biofuels in temperate countries has not prevented the European Union from fixing a mandatory level of incorporation of biofuels into road fuels of 5.7 per cent for 2010 and 10 per cent for 2020, a decision taken, like all the other ecological objectives of European Union, without any economic or feasibility studies. These rates of incorporation have no chance of being reached and are all the more questionable since we are all conscious of the consequences of the competition between producing food and producing energy.

As is the case for many others, the main consequence of these European decisions will, in the end, be financial in nature. Having attempted unthinkingly to follow injunctions received from Brussels, France has decided to heavily subsidize the country's production of bioethanol and the corresponding cost exceeds a budget of a billion euros a year, to add to the long litany of needless expenses agreed to with the illusory aim of 'saving the planet'. It was only in late 2008, that Brussels, like Paris, reviewed its position, thereby recognizing the mistakes made.

The second generation

The plants now used to make biofuels were not selected by our ancestors to propel vehicles. Whether cereals, sunflowers, rape or beetroot, they were chosen for human consumption and have in no way been optimized to produce fuels, which helps explain the poor yields most of them produce when used to this end.

It would perhaps be a different matter if the plants had been selected by virtue of their capacity to provide massive amounts of petrol substitute and this is why a great deal of research is now underway to develop or even create plants capable of producing large quantities of 'second generation' biofuels.

Companies that have invested in this field can be counted in tens if not hundreds, and are to be found on every continent. Whilst oil prices were stable at tens of dollars per barrel, the game wasn't worth the candle. But it is a completely different matter when they approach, or even more so, exceed, a hundred dollars. The first company to find plants and processes that can be used to mass produce biofuels at a price in the range of one or two hundred dollars the barrel is sure to make a fortune, if not immediately, at least in the medium-term.

Three avenues of research are currently being explored with regard to three basic lines of products: grasses, trees and seaweed. The first two families of vegetable matter would provide ethanol suitable for petrol engines, and production of the latter would be suitable for diesel engines.

Four herbaceous plants are currently the subject of most of the research: sorghum, sugar cane, switchgrass – a tall grass from the American prairies – and, miscanthus, commonly known as 'elephant grass'. As for trees, hopes are pinned on the eucalyptus, the poplar and *radiata* and *taeda* (loblolly) pines with their exceptionally rapid growth.

As for the technical procedures that could be used to extract biofuel from these plants without incurring expenses and energy consumption that would cancel out the benefits of the operation, several ways are being explored. Some involve thermochemistry and consist of transforming the entire plant into gas or liquid using the heat of the biomass. Others rely on biochemistry, that is to say enzymes, and aim to split the components of the plant into glucose.

No one can say today if the research will bear fruit, and if so, when. But the advantage of developing high-yield plants and procedures is two-fold: they would take over from traditional petrols, yet would not rival the agricultural crops used for food. A study by the American Ministry of Energy and Agriculture[18] has, for example, emphasized that even with progress that is limited in comparison to the current situation, *it should be possible to grow 1.3 billion tonnes of plants for the production of energy without affecting food production, and to produce 350 billion litres of biofuels that would cover 65 per cent of the current consumption of petrol products... and even 100 per cent if the Americans would agree to buy more modest cars.*

Even if this doesn't happen tomorrow, there is no reason why long-term prospects should be different for Europe if research is successful. With biofuels, we are only at the very beginning of a new page in the world's energy history and it is reasonable to be optimistic in this regard. It is true that when petrol is sold at 1.50 euros a litre at the pump, which corresponds to 300 dollars the barrel, European motorists continue to buy it, albeit unwillingly, so there are reasons to hope researchers will one day develop profitable second-generation biofuels.

If second-generation biofuels can provide yields of five tonnes of 'petrol equivalent' per hectare, which is the aim of current research, five million hectares, that is to say less than 10 per cent of France, would suffice to meet the future needs of this country. Admittedly, this still represents considerable amounts of land, but it is not totally unthinkable, especially if the operation could be carried out on land that is not currently allocated to food crops.

The future is now

But it could be we are quite simply looking in the wrong place and the solution is already in our hands. The potential of biofuels is very different according to whether they come from the tropics or from countries with temperate climates. Sugar cane is already producing yields of four tonnes of 'petrol equivalent' to the hectare, compared to one painfully produced tonne for beetroot. When it comes to oils, the difference is even greater. Oil palms sometimes produce seven tonnes to the hectare, which is about ten times more than our rape. The Sahel, for its part, is beginning large-scale cultivation of a plant known as *jatropha*, which can produce two tonnes to the hectare on land that is arid and unsuitable for any other crop. To sum up, and as might have been expected, yields from sun-soaked countries and temperate ones are incomparable.

This is why, faced with the inevitable drying up of underground oil someday in the future, the solution to its replacement may already exist with sugar cane and palm oil as the principal elements. Brazil already produces 25 million tonnes of ethanol from sugar cane, the equivalent of 19 million tonnes of petrol and it does so on seven million hectares, that is to say 70,000 square kilometres. Moreover, without even touching the Amazonian forest, Brazil has almost 100 million hectares of non-exploited land to which may be added 60 other million hectares currently used as extensive grazing land. With a yield of four tonnes of 'petrol equivalent' to the hectare, likely to be improved by the genetic research that Brazil has specialized in, it is easy to see that the country's production could, if need be, reach or exceed the level of Saudi Arabia, which is producing approximately 500 million tonnes of oil a year.

On the other side of the world, palm oil is king. Indonesia and Malaysia are in the process of transforming their equatorial forests into gigantic oil palm plantations. The surface area of these has already reached 4.3 million hectares in Malaysia and it is planning to expand this to 8 million hectares, that is to say 80,000 square kilometres, which would then represent almost a quarter of a country. If the country adopted this policy, it is because it becomes profitable financially as soon as the price of the barrel soars. When these plantations are producing at full capacity, they will be providing not far from 50 million tonnes of palm oil, a product that can be transformed to fuel diesel engines. By way of a historical footnote, in 1890, Rudolf Diesel ran the first engines to bear his name on vegetable oils and as early as 1890 predicted that 'the use of vegetable oils for engines may seem negligible today. But these fuels may be as important in the future as oil and coal are today'.

Viewed from a planetary standpoint, the replacement of petrol does not therefore seem to be an insurmountable task. We have 135 million square kilometres of land mass and if 10 per cent of these could be devoted to the production of fuel with an average yield of five tonnes of 'petrol equivalent' to the hectare, a simple calculation shows that the world's annual production would exceed six billion tonnes a year, that is to say half as much again as that of 'natural' oil today. It may be worth mentioning in passing that French Guyana alone, with its 90,000 square kilometres could ensure France's self-sufficiency in fuel.

Nobody can say whether the allocation of currently uncultivated tropical and equatorial land will provide the solution when oil supplies have run out. Perhaps this is not desirable. But it is reassuring to know that we have, at least, a solution to hand. It is no accident that the billionaire George Soros decided to invest a billion dollars in the production of ethanol from sugar cane in Brazil, and that the biggest biodiesel factory in the world is being built in Singapore to transform palm oil produced in Malaysia and Indonesia into fuel, even though it is true that the fluctuating price of oil means it may not be profitable in the short term.

You should also not forget that the new techniques of drilling and operations that have recently been found will allow us to extract from underground very important quantities of unconventional natural gas and oil. These would come in addition to the present sources of these products, and would delay the date by which it will be necessary to find a substitute to fossil hydrocarbons.

The old spectre of the depletion of resources of the globe relies on a fundamentally erroneous analysis of things, because it does not take into account something essential, human inventiveness. The resources of the planet are considerable and will be always sufficient to meet the needs of humanity, either for food, ores, or energy. Should one of these resources come one day to be exhausted, progress will know how to substitute it with others.

II – Electricity for all
The major preoccupation of a quarter of humanity, that is to say 1.6 billion human beings, is not whether there will be enough fuel in

the future for vehicles, but quite simply knowing when they will have access to electricity. Others, even more numerous, only have access to limited, or sporadic electric current. For us, the inhabitants of developed countries, the benefits of electricity are part of our natural heritage, taken for granted in the same way as the water we drink or the air we breathe. We have forgotten how revolutionary it was to be able to have light in the middle of the night, cook food and run the numerous pieces of equipment that simplify daily tasks, and more recently, to be able to communicate with the whole world thanks to advances in information technology and, last but not least, the difference the arrival of air conditioning made to life in tropical countries.

As everyone knows, abundant electricity is, in addition, indispensable to all industrial activity. Did Lenin not declare, 'Communism is the Soviets plus electricity'? Without electricity, economic development is impossible, the quality of life extremely mediocre and health itself comes under threat. According to the World Bank, 500,000 Indians, most often women and children, die each year of cancers because they live in one room where they are permanently breathing in smoke from wood or coal fires used both as a source of heat and to cook food. It is therefore understandable that developing countries place the mass production of electricity and its distribution throughout the country at the top of their list of priorities.

This is what China has understood and acted on in recent years when it embarked on the gigantic programme that has enabled its recent development to take place. The figures are dizzying. The People's Republic has adopted a plan aiming to bring into service *each week for 20 years a 1,000 megawatt coal power station*, that is to say power worthy of a nuclear plant unit, so that *each year*, it has an additional 52,000 megawatts, the equivalent of 40 nuclear plant units of 1,300 megawatts. In reality, it has done more since the additional power created was 50,000 megawatts in 2004, 70,000 in 2005, and 102,000 in 2006. That is to say, more in the course of that one year than all the power installed in the last half century in a country like France that has a set of power stations that serve as a reference for Europe!

Under these conditions, we should not be surprised that consumption of coal in China doubled from 2000 to 2006 and now represents 40 per cent of that of the planet. We should not be surprised either that the result has been in line with the effort expounded and that the electrification of China has been declared an outstanding success story by the International Energy Agency, a success that has no equivalent to date and probably never will have[19]. In 1990, the majority of the Chinese still lived in homes with no electricity. In 2005, the connection rate had reached 99 per cent, though it is true that the power available was sometimes still modest. By way of comparison, the connection rate has only reached 62 per cent in India where the number of people without electricity exceeds 400 million. It has to be said that the coherence of China's development policy stems in large part from the fact that all its leaders are engineers, beginning with the president of the Republic and the prime minister. One of the remarkable characteristics of what constitutes, with the construction of its motorway network, one of the two pillars of recent Chinese expansion is that this gigantic effort has been achieved mainly without help from abroad. Power stations, turbines, high and low voltage networks, electric appliances, various engines, even the light bulbs themselves – everything or almost everything was made in China.

This unprecedented effort has relied almost exclusively on coal since more than 85 per cent of Chinese electricity is provided by coal-fuelled power stations with the consequences we have seen for world emissions of CO_2. More than half of the growth in planetary emissions of CO_2 recorded over the last ten years can be attributed to Chinese coal-fuelled power stations alone!

One question comes to mind therefore. Why do China, India, Indonesia, Vietnam and the others all rely on such a polluting form of energy to produce their current? The answer is simple: they have no choice.

Admittedly, dams can produce electricity without discharging into the atmosphere at all. But potential sites are limited. The famous Three Gorges dam itself, though the most powerful in the world to date, only provides a few per cent of the electricity China needs.

Nuclear power requires at least double the investment of coal-fired power stations and triple that of gas turbines. In addition, construction times are twice as long and after 30 years of interruption, a new generation of nuclear power stations is only now beginning to painfully emerge.

If we consider too that China and India have relatively abundant proven coal reserves with respectively 114 and 56 billion tonnes out of a world total of 850 billion according to official sources, it is understandable that these two countries, and many others, have counted exclusively, or almost, on this easily available and relatively cheap resource to produce the electricity that is indispensable to develop their economy and save their population from poverty.

A word of caution is needed, however. Reserves of coal may not last very long at the rate at which these countries' consumption is developing. China and India have become importers, mostly from Australia and Indonesia, and in the summer of 2008 had their first difficulties in obtaining supplies, with stocks at their lowest.

Is it possible to envisage other solutions for the future, therefore?

The mirage of alternative solutions
Leaving aside hydraulic power, other renewable energy sources often give rise to a great deal of hope. But they are subject to such limitations that they can only play a very restricted role at the moment.

As already mentioned, wind farms suffer from such limitations that they can only usefully operate for around a quarter of the time, that is to say when the wind is blowing sufficiently, but not too strongly. This is why they must be 'supplemented' the other three quarters of the time by another source of electricity production, which significantly reduces their usefulness. To remedy this major handicap, we would need to have technical solutions that allow us to store directly or indirectly, the electricity produced when the wind is blowing so as to be able to use it when it is not. In Quebec, for example, when the windmills are turning, the production of electricity produced by the country's many dams is simply slowed down and the water saved for periods of no wind. We can imagine a distant future where other countries would equip themselves on a large scale with artificial reservoirs where sea or river water would be sent up when the wind is blowing so that the energy could be retrieved as needed. At present, however, this is a dream rather than a reality, for the cost would be prohibitive.

As for solar power, everything depends on its future cost price. At present, it is extremely high and existing technical solutions are only competitive for isolated installations that would be too costly to connect to a general power grid.

Over and above solutions that are at present marginal, only nuclear power would seem to be a certainty for the mass production of current. This is why China has ordered several reactors of each type from Western and Russian suppliers in order to master the technology with a view to copying it later, because it represents an almost inexhaustible source of energy. It should be noted that current reactors only use energy from fission induced by thermal neutrons (slow neutrons) and use mainly uranium 235, which means they are only retrieving around 0.6 per cent of the energy potentially contained in natural uranium. In the longer term, fast neutron reactors using uranium 238 will see the day (generation IV). Operating as 'breeder reactors' they will retrieve 70 to 90 per cent of the energy of the initial uranium instead of 0.6 per cent. *As a result, uranium resources will therefore be multiplied by a factor of at least 50 compared to now.* The investment needed for such fast neutron reactors (FNR), will doubtless be higher than for one of today's pressurized water reactors but the cost of the kWh produced will then be independent of the price of the natural uranium. But many technical problems remain to be studied and the reliability of the option must be demonstrated before large-scale industrial use can be implemented, probably around the mid-twenty-first century. However, when this has been done, uranium will then become an almost inexhaustible resource on the human scale, and take the place of coal.

The opposition of ecologists to nuclear power is all the more puzzling therefore since they also reject coal-fired power-stations, when these are the only two realistic solutions for producing electricity on a massive scale.

III – The great climate fear

There remains one other early twenty-first-century fear haunting the minds of most of the people since it is ceaselessly being rehashed. Several nagging questions serve to sum it up. Will the climate change? What will the consequences be for us and for the other inhabitants of the planet? What should we do about it?

As we have seen, the answer to the first of these questions is not in doubt. Yes, the climate will change over the years to come as it has always done on Earth as on other planets, with periods of cooling and warming succeeding each other, though climatologists cannot at the moment agree on the causes of these fluctuations. As we have also seen, it is very unlikely that CO_2 is a sizable contributor since there is no visible link between the evolution of its concentration in the atmosphere and fluctuations in the average temperature of the globe. Nothing can therefore be excluded with regard to the future, not even the hypothesis of new periods of cooling since the temperature of the planet has not risen since the year 2000. We must therefore above all remain measured in our response, and those who make alarming predictions concerning the evolution of the climate by the end of the twenty-first century should not be taken seriously. What is more, even if they were to come true, humanity would be able to cope by exploiting the possibilities of technological progress.

It is the same story with the level of the sea, which, the IPCC tells us, could rise by 30 centimetres in a century. It is evident that if this prediction were to come true, we would have nothing to fear. The level of the sea has always varied and in far greater proportions. At the time of the Crusades, Saint Louis set off for Palestine from Aigües Mortes, a town whose present-day name clearly indicates that it is now located inland. If we go back several thousand years, the situation was still more different. Twelve thousand years ago, the English Channel did not exist. Great Britain and France were joined by an isthmus that would have changed the course of the history of the world had it subsisted. What incredible upheaval, in which Man clearly played no part, must have happened 12,000 years ago for the ice that then covered the northern half of Europe and America to melt and the level of the sea to rise by more than *100 metres in just a few centuries*? Alongside this, our 'warming' is just a little joke. At the height of the last ice age, did not the ice cap cover 28 million square kilometres, that is to say 14 times the surface area of Greenland today? It's not going to melt twice...

Numerous proofs of wide fluctuations in sea levels exist all over the world compared to which an elevation of 30 centimetres is insignificant. If this happens, it will suffice to elevate existing dykes by the same amount or to build them if they do not already exist. Whatever the circumstances, the creation of dykes to protect the coastal zones situated at around sea level will be essential in many places in the world. The Netherlands, part of which is below the level of the North Sea has long shown how to make threatened regions safe, and thanks to modern techniques, it is a solution that is beginning to be implemented in other places such as Bangladesh and the Maldives who are expanding their country as a consequence. Whether it will be necessary to build to protect zones threatened by tides, storms and cyclones 30 centimetres higher or lower, is neither here nor there.

As we mentioned in Chapter I, it is only with the hypothesis that sea levels would rise by several metres, put forward by people such as James Hansen, that things would become rather more serious. But it is a hypothesis that has not been accepted by the

IPCC experts who are usually inclined to pessimism. Even if this were to happen, we have every reason to think that humanity would be able to gradually meet the challenge by the end of the century as and when the need occurred through the implementation of civil engineering programmes aimed at protecting all or part of the world's low-lying zones, starting with the towns that are situated there, as the Dutch have done. But, it has to be emphasized; the vast majority of scientists have rejected this extreme hypothesis.

On the other hand, two other consequences often evoked with regard to the change in the composition of the atmosphere deserve our attention: the consequences of a possible rise in temperatures and the possible growth in the number and seriousness of cyclones, hurricanes and other exceptional climatic events.

The myth of climate refugees
As everyone knows, the models constructed by the IPCC experts under the aegis of the United Nations indicate that the average temperature of the globe could increase by two to three degrees celsius over the present century. Although this is not at all certain since these models are based on postulates, it is a hypothesis that deserves to be taken into consideration. Would this result in a catastrophe, as many people would have us believe?

The least we can say is that it is not at all clear as the human race's capacity for adaptation is remarkable, unlike that of other members of the animal and vegetable kingdom. People have been living in Abidjan (average temperature 29 degrees) and Stockholm (average temperature 6 degrees) for a long time now, not to mention the Eskimos and other Inuits who endure even tougher climates. Within one and the same country, differences can be considerable. Chicago is, on average, colder than Houston by ten or so degrees. The climate in the foothills of the Himalayas is very different from that in southern India. Even in France as the film comedy *Bienvenue chez les Ch'tis* (*Welcome to the Sticks*) shows, the climate is not the same on the Mediterranean coast as it is by the North Sea and there is more than a few degrees difference in average temperatures between the two extremities of France.

Nonetheless, no-one has ever heard tell of the inhabitants of the southern regions of America, India and France invading the north of their respective countries or the reverse. When migrations occur, they have completely different causes. The invented idea of 'climate refugees' is nonetheless present in many official reports, beginning with those of the United Nations. If we are to believe them, the developed countries will soon be threatened by the arrival of hundreds of millions of 'climate refugees', as though it were possible to classify those who are seeking to flee poverty into two categories: climate refugees and others.

A difference in climate does not cause the flow of Mexicans towards the United States and Africans or Asians to Europe, but under-development and poverty. And it is fighting poverty, whatever climate changes may occur, that will cause these migrations to eventually dry up.

Fortunately, economic progress in developing countries shows that the elimination of poverty is not an inaccessible objective for the future, despite the inevitable stops and starts, even if it is clear that it will be longer and more difficult to achieve in certain parts of the globe than in others, and in particular when and where the birth rate is not under control.

It should be added that the developed countries, having understood that they could not 'open their doors to the entire world's poverty', are gradually setting up increasingly effective mechanisms to control illegal immigration that make the idea of being invaded by uncontrollable streams of 'hundreds of millions of climate refugees' still more unlikely. In particular, this is what the European Union, at the instigation of France and others, has just done.

Most of those who inhabit the underprivileged regions of the planet will therefore remain where they are and Man, as ever, will adapt to future temperatures, whether they are higher or lower than in the past. And it is to combating poverty and encouraging the development of the poorest that the rich countries should allocate

Evolution of tropical cyclone activity

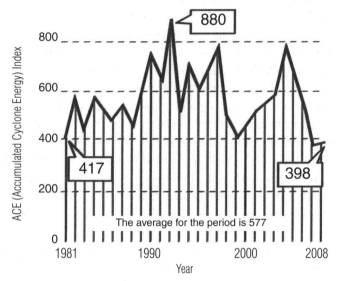

Fig. G The ACE Index from the University of Florida's Center for Ocean-Atmospheric Prediction Studies (COAPS) is a measure of the number, strength and duration of all tropical storms in a given year.

the huge sums they are now wasting with the illusory aim of 'saving the planet'.

Cyclones and tornadoes

Nonetheless, the possible consequences of changes in the atmosphere are not limited to rises in temperatures. Must we, as many fear, expect an increase in atmospheric phenomena such as the storms, tornadoes and cyclones that periodically ravage certain areas of the planet?

On this point, the IPCC has observed that tropical cyclones are no more frequent than before '*but could be more serious*'. In this field as in many others, the facts happily do not confirm these declarations, as the benchmark records held by the University of Florida attest (see Fig. G). Storms, tornadoes and cyclones have always occurred and are nothing new. Historical records tell of a hurricane of extreme violence which left 8,000 dead in its wake in England in 1703. In 1881, a cyclone ravaged Vietnam leaving 300,000 victims. India has been hit several times over the centuries with tens of thousands of deaths each time. And there has been worse. In 1932, diluvial rains caused China's three great rivers to burst their banks and the death toll was estimated at three million, with a more accurate figure impossible to obtain. In recent times, we can all remember the dramatic events that hit Burma in 2008, with the death toll exceeding a hundred thousand.

Since 1981, the University of Florida has been following an index representing cyclone activity entitled, 'Accumulated Cyclone Energy' (ACE) that takes into account the number, strength and duration of all tropical storms, hurricanes, typhoons and cyclones occurring in the northern hemisphere.

The fluctuations of this index do not show any worsening of these occurrences. Quite the contrary. The years 2007 and 2008 were the most *inactive* ever recorded since the calculation of the index.

However, this lengthy list of cyclones no longer represents the threat to man it previously did, mainly due to two new circumstances, both the fruits of technical progress, which are currently reversing the order of things.

Almost every season, the media recite a litany of the tropical cyclones that cross the Gulf of Mexico and which are extremely violent some years. But the number of victims has never been so low. Whilst they used to be counted in tens of thousands, this is no longer the case, and not in this region alone. In 2006, a cyclone of unprecedented strength, similar to that of 1881, crossed the middle of Vietnam, causing considerable material damage. But there were practically no deaths and we heard nothing about it.

The reason for this is that meteorological satellites have changed everything. Warned several days in advance, populations can now take the necessary precautions, consolidate buildings, and above all, take shelter. They are all the more able to protect themselves since their standard of living makes it possible to implement measures that could not previously be envisaged. Villagers in Bangladesh, for example, have built artificial hills on which the population can take refuge if need be. The use of concrete has meant that buildings can be built that will withstand the most violent of winds and accompanying rains.

Things are at an even more advanced stage in the United States, where the size of the automobile fleet means hundreds of thousands, if not millions of people can be evacuated in a day or two. Loss of human life can now usually be counted on the fingers of a few hands. The only recent exception was Hurricane Katrina that resulted in more than a thousand victims in 2003. But if New Orleans was flooded, it was the incompetence of the local authorities that had not maintained the city's protective levees that were responsible and not the cyclone itself.

All over the world, victims of cyclones are becoming an exception when two conditions are fulfilled. The populations under threat must first of all be correctly warned. This may seem self-evident, and thankfully it is more and more often the case in the age of television, radio and perhaps even more so, of the mobile phone. If the 2008 event in Burma had such terrible consequences, it is because this country has one of the most backward dictatorships of our time which did nothing to warn its people of the strength and seriousness of the cyclone that was about to hit them. Casualties would have been minor if it had been a democracy, or properly organized, as China is today.

But, even if it is essential, information alone is not enough as in the case of Haiti which was was hit three times in 2008 by devastating cyclones leaving hundreds of dead. Without even mentioning the awful 2010 earthquake, the extreme poverty of this underprivileged country has prevented it from taking the protective measures implemented by the other Caribbean islands that have allowed them to face the extreme weather conditions, which have always hit the Gulf of Mexico, without almost any loss of life.

In short, the way things are going is no longer in doubt. Cyclones, tornadoes and other tropical hurricanes will, in future, no longer claim victims or very few, and major human catastrophes will, from this point of view, become increasingly rare, even though material costs are set to increase.

From this point of view, there is no reason for gloom about the future. In its various forms, progress will continue and we should be devoting ourselves to accelerating and diffusing it. And we should not forget that in a distant past, notably during diplodocus' era, CO_2 concentration was five to twenty times higher than today, with no known adverse consequences.

Prevention or adaptation?

What is to be done therefore? Since it is illusory to count on modifying the CO_2 emissions that result from human activities and even less the planet's possible climate changes, human solidarity dictates that we should be seeking to limit the consequences of these, should they occur. No one can say in fact whether or not sizable climate changes will take place in the course of the twenty-first century, since they have always existed. In other words, since the *prevention* of phenomena that we cannot control is impossible, let us help those who may be the victims of them to arm themselves through policies of *adaptation* to the possible risks. In the face of drought, let us build wells, develop irrigation and seeds adapted to the lack of water. Faced with flood risks, let us build dams. Faced with the dangers of cyclones, let us develop public warning systems and build refuges.

The cost of a coherent policy of adaptation to climate risks has been evaluated at a global level at tens of billions of dollars a year by the World Bank and the United Nations, with average estimates of around 50 billion[20]. Such a sum obviously seems considerable. But, if we think about it, it is much less than the sums of billions now being spent at pure loss by the rich countries with the illusion of reducing CO_2 planetary emissions. Nobody, however, imagines that such a large sum of money can be released at once, and it is only progressively that it can be constituted as a global fund for adapting to climate changes, whether these are human in origin or not.

A first fund of this kind was recently created by the United Nations, but was allocated just 300 million dollars, only 10 per cent of which has been spent, a derisory figure if ever there was one. Faced with such a situation, China suggested for its part that the rich countries attribute 0.5 per cent of their GDP to the risks which may result from climate changes, which would represent a hundred billion dollars a year.

Let us be under no illusion, however. Such a sum is unimaginable, if only because it would be difficult, if not impossible, to identify from among initiatives to combat poverty and in favour of development, those whose aim is specifically adaptation to climate change. Most of the expenses relating to 'adaptation to climate risks' should be directed towards what the countries concerned should do whatever happens, whether there is climate change or not: irrigation, the construction of dykes, the development of seeds that withstand drought, etc.

In other words, in the end, devoting money to adapting to possible climate changes comes down to increasing aid to developing countries, and we know the extent to which this one is parsimonious. The paradox is that the rich countries are capable of wasting hundreds of billions of dollars a year in the name of a hypothetical reduction in CO_2 of human origin, yet we should not cultivate the illusion that they will be willing to release significant, yet far lesser sums, to help poor countries adapt to the consequences of possible climate changes. It is only at first sight that this is a paradox. In the first case, the spending takes place in the countries that are financing it and where powerful lobbies with vested interests have persuaded public opinion and their leaders that they should act in order to 'save the planet', even though the only way to save, not the planet, but those who live on it and their descendents, would be to help the poorest of its inhabitants and their children emerge from poverty, whether or not we do this in the name of the fight against the risks of climate change or as part of a general move in favour of the third world.

What goes for countries is also true of individuals. Most of those who want to take action to improve the future of the world are sincere and really do want to help its current and future inhabitants enjoy a better life. If this really is their aim, there are many ways in which they can act, in particular by helping the many foundations that devote themselves to improving the lives of the most deprived. At his own level, this is what Bill Gates does. He is not seeking to save 'the planet', but the children who live on it. His immense fortune, reinforced by that of Warren Buffett, is largely devoted to the widespread vaccination of third-world children, with amazing results. If anyone deserves a Nobel Peace Prize it is without a doubt the American billionaire and not the impostor from the same country who recently received it.

CONCLUSION

From Rio to Copenhagen

With hindsight, it will be hard to believe the facts described in this book.

Their origins are to be found in the 1987 report, known as the Brundtland Report, of the World Commission on Environment, which coined the notion of 'sustainable development'.

However, it was the 1992 Earth Summit in Rio, at which 40,000 people gathered, including participants from hundreds of NGOs, that marked the beginning of the strange times through which we have just lived. A handful of people, the most prominent of whom was the Summit's organizer and secretary-general, Canadian billionaire environmentalist Maurice Strong, had just sealed an alliance between the United Nations and environmental movements, and launched what was to become a quasi-religion.

This new messianism claimed that man was endangering the planet by upsetting its balance and depleting its resources. This reproach was not new and had already been expressed long before then. What was new, however, was the claim used to support the accusation, i.e. that greenhouse gas emissions resulting from human activities, notably CO_2, were disrupting the climate and causing global warming, with catastrophic, irreparable consequences for our descendants.

For those who had had a negative view of mankind for generations, the missing 'proof' had finally been found. And it was provided by a handful of climatologists who shared the same view of things and claimed that the planet was heading for disaster as a result of human activity.

It was at this time that the Intergovernmental Panel on Climate Change (IPCC) was formed, the very statutes of which affirmed mankind's responsibility, given that the body's mission was to examine 'human-induced' climate change.

As the first head of this new organization, Sir John Houghton was said to have declared at the time that: '*unless we announce disasters, no one will listen*'. Very late on, he denied the truth of this statement. Right or not, this approach was to be an all-out success, fuelled by announcements of ever-more alarming disasters, such as rising ocean levels that would submerge entire regions, the melting of the Himalayan glaciers, the disappearance through drought of the Amazonian rainforest, growing numbers of cyclones and hurricanes, expanding deserts, the extinction of the polar bear and many other species, the invasion of rich countries by hordes of 'climate refugees', and numerous other disasters.

All, or almost all of this subsequently turned out to be untrue since it had not been based on any scientific research, contrary to what was widely maintained at the time. But such predictions were issued year after year by the IPCC whose control the founders had been careful to monopolise. For almost two decades nobody, or almost nobody, dared to challenge these apocalyptic predictions and their most tangible implication: that, at all costs, we had to reduce global emissions of their major cause, i.e. CO_2.

I must admit that, for a long time, I also subscribed to this way of thinking. How could I have imagined that a United Nations-affiliated body that claimed to have the support of thousands of experts, that was awarded the Nobel Peace Prize and that was backed, into the bargain, by the world's heads of state and their hundreds of advisers, could tell anything but the truth? So for a previous book, I accepted the IPCC's assertions as truth.

But my eyes were opened when I focused my attention on rising sea levels and realized that the predicted consequences were totally unfounded. How could a rise of three centimetres a *decade* wipe entire regions off the map?

Those who affirmed this had obviously confused centimetres with metres, just as they would later confuse the year 2035 with 2350 for the melting of the Himalayan glaciers. I had to face facts: those who produced the IPCC's reference document, its 'Summary for Policymakers', were not credible. What they claimed defied common sense and was systematically biased in order to create panic.

But, of course, I was not the only one to have been misled. The blindness that engulfed the entire planet, including all the world's leaders, was so intense that nobody asked the elementary question that is the primary focus of this book: *'Is it realistic to imagine for a split second that we can increasingly use the hydrocarbons that are needed for life and the development of mankind, i.e. oil, natural gas and coal, whilst simultaneously restricting the emissions that result from their use?'*

The prospect of future catastrophes prevented this question from even being asked. In the face of such a dire future, there *had to be* solutions to reconcile human development and control of the emissions that were endangering the planet. That this was physically impossible was quite simply inconceivable. Among the United Nations' thousands of experts and advisers, as well as governments throughout the world, nobody uttered the obvious truth, i.e. that it is impossible to pursue the development of mankind, which will for a long time to come need to use hydrocarbons, and at the same time reduce the CO_2 emissions that are an inevitable consequence of their use.

So the most basic common sense vanished for nearly 20 years until the nonsensical summit in Copenhagen, the stated objective of which was to halve global emissions of CO_2 when they were, in fact, set to *double* due to the development of emerging countries. In the interval between the Rio and the Copenhagen summits, emissions had already increased by 50 per cent, rising from 20 to 30 billion tonnes a year, and they could only continue to grow in the future bearing in mind what the poor countries that comprise the bulk of mankind are currently doing or planning to do.

Upon reflection, the issue involved was one of such importance and with such serious consequences that there is something deeply worrying about this global blindness. It should be seen as a consequence of the democracy that Winston Churchill referred to as 'the worst of all systems, apart from all the others'. If they are to be elected, politicians cannot go against public opinion. And when this falls victim to global manipulation, organized moreover under cover of the United Nations, democracy is in danger since the leaders themselves can no longer choose the viewpoints that they express.

There are many Western politicians among the 'sceptics' but it would be suicidal for any of them to admit this. Even heads of state fall victim to intellectual terrorism. It is common knowledge that the French president, Nicolas Sarkozy, wanted to bring Claude Allègre, an esteemed scientist and a leading sceptic in France, into his government. However, the prospect provoked fury among environmentalists, not only within the French Green Party, but also across the entire political chessboard, including the president's own party, the UMP. Faced with the prospect of having to fight a difficult battle on all fronts that may have paralysed all his policies, Nicolas Sarkozy had to abandon the idea. What alternative did he have in the face of French national public opinion, the vast majority of which was convinced that the planet is heading towards disaster and that man is responsible? So he chose to go along with this opinion. By the way, this choice did not help his approval ratings at all, but proved to do a lot for the credibility of the French Green Party (Europe Ecologie) which obtained very unexpected and high ratings at the following elections.

Will the truth come from the United Kingdom following the 2010 elections? Will it come from the United States, where the Republican Party has a natural tendency to subscribe to the arguments of the 'sceptics' in order to restrict government intervention? Will it come from Australia, New Zealand or elsewhere? Which country will be the first to admit it is impossible to reduce global CO_2 emissions? In the meantime, it is a non-democratic country that has moved things along, making Copenhagen a turning point from where there is no turning back. For indeed it was China that wrecked the Copenhagen summit by realistically refusing any targets for reducing global CO_2 emissions because it knew that it would be impossible to respect them. In early 2010, the country's representatives went even further by declaring for the first time that it was finally time to listen to the arguments of sceptical scientists with regard to the influence of humans on climate change, instead of vilifying them and systematically excluding them from any debate on the issue.

The reason why China was able to set itself apart in this way is because it is not a democracy. Its leaders do not have to follow public opinion and can, if necessary, distance themselves from it when they consider this to be in the interests of the country. Their background also differs significantly from that of Western leaders. The top leaders of China's Communist Party are all engineers and therefore unlikely to be swayed by emotional or irrational arguments, in contrast to politicians who are elected because they have a way with words, which is the case in most democracies.

It is paradoxical, to say the least, that in this particular instance it was a non-democratic country that did something for the good of mankind, but this is the case. For pointless spending, allocated under the illusion that it will help 'save the planet', is reaching such astronomical proportions that it will have a lasting, detrimental effect on those countries that indulge in it. In France alone, it is estimated that such spending will amount to 440 billion euros by 2020, virtually all of which has no economic justification and will, therefore, be a pure waste of money.

In this way, France, like other countries, has been misled by the United Nations' conclusions that affirm that it is possible to reduce the planet's CO_2 emissions. The conclusions emanate from an IPCC working group that attracts little attention. This is Working Group III, whose mission is to see how emissions can be mitigated. By its very nature, this working group does not fall within the scope of a single category of experts, since it needs to deal with fields as diverse as energy, transport, buildings, agriculture, forestry, waste management and so forth to assess how, and at what cost, it would be possible to reduce each of their emissions. So one can imagine the difficulties involved in putting together a summary, given the fact that nobody can specialize in such diverse, unrelated subjects at the same time.

This is borne out by examining Group III's conclusions, which are utterly incoherent. One can read, for example, that it should be possible to obtain a stabilization of emissions by 2030 at a cost of '20 to 80 dollars per tonne of CO_2', which is so imprecise that it ranges from a factor of one to four! But what is significantly more serious is that the values put forward are incomparably lower than the reality. As previously mentioned, it could be observed that, when the price of fuel at the forecourt reached 1.50 euros per litre in Europe in 2008, corresponding to almost 600 dollars per tonne of CO_2 emitted, this did nothing to stop European motorists from continuing to use their cars. The truth is that to have a massive effect on emissions, it would be necessary to allocate spending amounting to hundreds of dollars per tonne of CO_2. When it is borne in mind that current emissions in the atmosphere from hydrocarbons alone amount to 30 billion tonnes per year and are bound to increase, one can immediately see that the cost of significant action would be prohibitive and inaccessible. In other words, this confirms the fact that it is, in practice, impossible to do without oil, natural gas and coal and avoid the corresponding CO_2 emissions. But Working Group III, whose work, when examined, proves to be woefully vacuous has allowed the opposite to be believed by affirming that it is possible to control the planet's emissions at a reasonable cost. It is this erroneous affirmation that Sir Nicholas Stern made his own without checking and ascertaining its truth.

One can, therefore, ask oneself whether attention was not mistakenly focused on Working Group I, which is dedicated to studying climate change, rather than on the third group. Focusing on the latter would have shown that nothing can be done to limit emissions to a significant extent. It is true that, for the most part, the first group only involves a single profession, i.e. climatologists, whilst the third involves the participation of specialists from such diverse fields that nobody feels competent enough to judge their conclusions. Yet one merely needs to read some of the assertions of this working group to discover the extent to which they are utopian.

So, at the global level, hundreds of billions of dollars are now being wasted each year to 'save the planet'. For instance, the planet is gradually being covered with wind turbines that only exist because they are subsidized to an extent that is unheard of in any other area. Their construction is all the more unwarranted since they cannot be relied upon when they are needed, owing to the fact that they only work when it is windy, that is to say in an unpredictable way, which brings the value of the electricity they produce to very little. In addition, the power they provide is many times more expensive than the cost price of electricity from traditional sources that can be relied upon in all weather conditions, not just when the wind is blowing. According to a report from France's Commission of Energy Regulation, an official independent body, French consumers will see their bills needlessly rise by three billion euros a year if EU Directive targets are met. It should be added that wind turbines disfigure a landscape that has been passed down to us through generations and that is one of the greatest charms of the Old Continent. And, as the technology stands at the moment, electricity from photovoltaic sources is even more ruinously expensive, with a cost price that is between five and ten times higher even than electricity from wind-generated sources and which, once again, owes its survival to disproportionate subsidies. We are living in the illusion that power from the wind, sun and waves is free. It is nothing of the sort. In fact, the investment needed to capture this 'free' energy is presently hugely expensive, which explains why no such project would be viable without huge subsidies.

What is true for energy is just as true in other areas. Seeking to wean drivers away from their cars or to discourage companies from using lorries is extraordinarily expensive, in addition to being doomed to failure. We are living in an age when time is extremely valuable. Trying to replace a lorry journey that can be completed in a few hours with one that would take several days by train is, for the most part, mission impossible. It is no coincidence that more than 95 per cent of goods transportation by value, is carried out by lorry in Europe and that 90 per cent of people's non-city centre motorised journeys are by car. It is because the use of road vehicles is in the interests of the functioning of the economy and the well-being of the population. All the costly efforts that have been made to change this situation have failed.

For its part, the insulation of buildings is only justified when its cost does not exceed the anticipated benefits. If this is the case, the work in question must obviously be carried out. But, quite often, such work is only viable because it receives subsidies that are, once again, financed by taxpayers.

As far as this game is concerned, Europe is currently breaking all spending records in the name of ecology, without realizing that what it is doing can have no significant influence on global emissions and that it is, in fact, seriously and pointlessly hampering its future in the global competitive landscape. This way of operating also means that the world is unable to focus on the true priorities of our times, such as the fight against hunger. With a few exceptions, the world's heads of state were all in Copenhagen. But none attended the FAO World Summit that was held a few weeks earlier in Rome and at which it had to be noted that the few billion dollars that would suffice to feed several hundred million human beings properly were nowhere to be found, at a time when hundreds of billions are being spent under the illusion of 'saving the planet'.

What is now happening is that everyone wants to play a part in 'sustainable development'. Obviously, nobody can be against development, which is needed to lift the great majority of mankind out of poverty, or against the fact that it must be sustainable to enable future generations to reap the benefits too. The term should be received with unanimous approval were it not for the fact that, in reality, it implies something else. It implicitly gives the impression that our present development is not sustainable and that the very foundations of our civilization need to be changed.

This is not true. The debate over climate change has become as virulent as it is today because it is part of a much wider picture, that of man's responsibility with regard to the evolution of our world, which is uniformly presented as being appalling and bound to lead us to disaster if nothing changes.

This way of looking at things is kept going by countless books, films, articles and statements that paint an increasingly gloomy picture for mankind, creating pessimism among young people and predicting the apocalypse for future generations.

I absolutely do not share this way of looking at things, which carefully sifts through the facts to keep only those that ignore the main point. Between 1900 and 2000, in a single century, life expectancy in poor countries, that is to say four-fifths of humanity, almost tripled, increasing from 27 to 65 years. Behind these stark figures lies a historically unprecedented turnaround since most of this incredible progress is due to the decline in infant mortality. While two out of every three children used to die, nowadays this figure is well below one in ten.

So the figures take on a whole new meaning when one knows how to interpret them. They mean, quite simply, that over the course of the twentieth century, *more than a billion* mothers did not have to experience the grief of burying a baby, a little boy or a little girl. *More than a billion mothers, and as many fathers.* Do we who live in rich countries have a monopoly on maternal love and suffering?

And progress is continuing at a barely credible pace, notably thanks to Melinda and Bill Gates's foundation, which is working to help achieve widespread childhood vaccination, and so successfully so that measles will have disappeared from the face of the Earth in a few years' time. Since the year 2000, the number of children who die *each year* before the age of five has fallen even further, by four million!

So how can we claim to be living in a worsening world and to be inevitably leaving a planet with a dire future to our children who, incidentally, will live rather than die? I obviously do not wish any harm to plants, whales, birds or polar bears, which, by the way, have been multiplying rapidly since commercial hunting was banned, their numbers having increased from 8,000 to more than 20,000 over the past 20 years contrary to what one reads everywhere. But no one can convince me that there lie the most important issues compared with the grief of parents who lose their children.

Falling child mortality also creates a virtuous circle. Once women around the world realize that their children are no longer dying, there is a massive reduction in the number of births. This trend is, more often than not, dramatic. Three decades ago, women in Morocco had more than seven children on average. Today, this average stands at scarcely more than two. And women in Tunisia now have fewer children than women in France. This is a global trend. In contrast to the generally accepted idea, we are witnessing a demographic fall on a worldwide scale. Followers of Malthus who raise the spectre of an overcrowded planet, incapable of feeding its inhabitants, are at odds with the facts.

The only significant exception is still sub-Saharan Africa, but here again a trend towards controlling births is underway and it is gathering pace, beginning with the coastal countries. Indeed, it would be useless to hope to emerge from poverty if the birth rate remains at around six or seven children per woman once infant mortality has been reined in.

Progress is not just apparent from a demographic perspective; it also affects our environment, forests, rivers, the air we breathe and what we eat, with the impact on life expectancy that we are all aware of. Admittedly, environmental improvements primarily concern rich countries. The seas around France, which were dubious to say the least a few decades ago, have now been cleared of pollution and the waters have recovered their purity. The country's lakes and rivers have been cleaned up too. Salmon are swimming back upstream in the Loire and the Seine and 30 species of fish have been recorded in Paris, compared with just four or five 50 years ago. And France is, of course, not the only country where such results are being achieved.

Regarding local air pollution in the large cities of developed countries, progress has been as rapid as it has been unnoticed. Due to the depollution of factories, building heating and vehicles, the atmosphere is now far purer than it has been since the nineteenth century.

But all this is extremely expensive and the majority of poor countries, whose current priority is to lift their people out of extreme poverty, and the danger of illness and death, will not be able to afford this until some time in the future, as was once the case for the presently rich countries. But once their standard of living allows it, there is no reason why they should not follow the same path. Obviously, everything cannot be achieved in a day, or even a generation, although China, India, Indonesia, Brazil and many other countries are experiencing change at a rate that nobody could have believed possible and they too are beginning to take the appropriate action to improve their local environment.

Furthermore, the planet's resources are not going to be depleted in a hurry. And when one of them does disappear, as will one day happen to oil, human ingenuity will replace it with others. So, even though much still remains to be done, why is there such constant talk about what is going wrong, and never about what is going well, the latter having the edge by far?

Proposal for a Statement about Climate Change

POINT I
On the global level, we can do nothing significant about CO_2 emissions and concentrations.

POINT II
There is no proof CO_2 emissions and concentrations are or will be a significant problem for the planet.

POINT III
We have to stop wasting public and private money in the illusion it will 'save the planet'. Huge savings are at hand.

* * *

POINT I

- The question of CO_2 is no longer with the developed world. Today, rich countries are responsible for less than half CO_2 worldwide emissions and the proportion is declining rapidly. The question lies with emerging countries where the large majority of mankind is living and whose emissions are increasing rapidly and will continue to do so for they need energy to get their people out of poverty.

- In the foreseeable future, only coal, oil and gas can provide them the bulk of the energy they need, and it is unrealistic to think mankind will leave unused underground these resources. The coal, oil and gas rich countries will not use will be used sooner or later by developing countries, and overall global CO_2 emissions will continue to rise almost unabated until underground resources are progressively depleted.

- There is no realistic way to avoid these emissions reaching the atmosphere. Technical processes such as CCS are far too expensive to play a significant role, even in the most optimistic forecasts regarding their costs.

- We have to realize CO_2 concentration in the atmosphere will continue to increase and may double in the coming century. There is nothing we can do about it.

POINT II

- It is wrong to say the scientific community is unanimous about the possible influence of manmade emissions on the climate. This community is highly divided, scores of scientists disagreeing with the majority of official views that CO_2 and other greenhouse gases emissions caused by human activities may have a sizeable influence on the earth's climate, which has always been changing over time.

- Earth temperature has ceased to increase for the last ten years, contrary to all models forecasts. The truth is that present science does not allow us to predict the evolution of the climate.

- There is no proof that the present average temperature is especially high. Around year 1000, Greenland was green, and had been inhabited by European settlers a number of centuries.

- The present rise in sea level is close to insignificant – around 2 to 3 millimetres per year – and will not be of a different magnitude during this century, according to the official IPCC forecasts themselves. As a consequence, there is no danger whatsoever for any island or any sizable inland area to be flooded by rising seas in any foreseeable future. Those who say the opposite are just misleading people. Centimetres should not be confused with metres.

- The IPCC is thus misleading world opinion and the leaders themselves, creating without reason a highly negative vision of the future of our planet and of mankind. The consequences of such a behaviour are dire. Huge human and financial resources are distracted from the fight against poverty and misfortune the world over.

Point III

- Fossil energies are precious. They should not be wasted and should be saved as much as possible every time it is economically justified.
- But huge amounts of public and private money are presently wasted in order to try to reduce emissions with the illusion this will 'save the planet'. They amount in most developed countries to a sizable part of GDP and prevent economic recovery, life level improvement, and return to proper budget balancing.
- Subsidizing for this motive such excessive practices at high cost for the taxpayer and the consumer should be ended, either in the fields of alternative energies, of transport, of building insulation, or in any other one.
- On the other hand, research in these fields has to be strongly developed to prepare new solutions for the time when fossil resources will come to be exhausted.
- But time-scales should not be confused. This will not happen for a number of decades. Acting as if the problem were immediate and spending money today on uneconomic and often soon to be obsolete technical solutions is unjustified.

- As there is no proof human activities are having a significant impact on climate, there are moreover no mitigation actions to be undertaken for this purpose.
- Huge savings are easily possible right now by stopping useless expenses. Public and private money should be used to better purposes to help people and countries in need.

Notes

The author makes no apology for not cluttering this work with unnecessary references. His sources are for the most part readily accessible in the public domain, available on the internet or in easily available published documents. Only where this is not the case, has he provided the relevant endnote.

1.   Key World Energy Statistics, International Energy Agency.
2.   International Energy Agency, op. cit.
3.   British Petroleum, op.cit.
4.   Cf. my book « *L'énergie à revendre* » (J.C. Lattès publisher)
5.   *L'imposture. Pourquoi l'éolien est un danger pour la France.*(Editions du Toucan)
6.   Serge Galam, *Les scientifiques ont perdu le nord*, Plon.
7.   *Le Monde Diplomatique*, Atlas Environnement.
8.   Claude Allègre, *Ma vérité sur la planète*. Plon.
9.   Giles Sparrow, *Planètes*, Hachette.
10. United Nations Development Programme. Human Development Report. Oxford University Press.
11. *Le Monde*, 5 May 2007.
12. On this subject, see David Henderson in *Climate Change Policy, Challenging the Activists*, Institute of Economic Affairs, London.
13. See in particular *Le Climat: jeu dangereux, dernières nouvelles de la planète*, Jean Jouzel, Anne Debroise, Dunod, 2007.
14. *An Appeal to Reason*, Nigel Lawson.
15. Op. cit.
16. IPSOS/FFAC, July 2006.
17. Alexandre Rojey: Energie et climat: comment réussir la transition énergétique, Editions Techniques.
18. *The Economist*, 21 June 2008.
19   World Energy Outlook 2007.
20   *The Economist*, 13 September 2008.

Index